SARAH

THE TROPICAL AGRICULTL

Series Editor
René Coste
Formerly President of the IRCC

General Editor, livestock volumes
Anthony J. Smith
Centre for Tropical Veterinary Medicine
University of Edinburgh

Dairying

Richard W. Matthewman
Centre for Tropical Veterinary Medicine,
University of Edinburgh,
Scotland, UK

with the collaboration of
Dr Noël Chabeuf
Institut d'Elevage et de Médecine
Vétérinaire des Pays Tropicaux,
Maisons-Alfort, France

Available legumes:
Stylor
Desmodium
Lab-Lab
Centro.
Napier
Leucaena

450 kg Cow
250 kg heifer ①
180 kg ②
30 kg ③

150th YEAR
MACMILLAN

This edition first published 1993

The Tropical Agriculturalist Series originated under
the title *Le Technicien d'Agriculture Tropicale* published
by G. P. Maisonneuve et Larose, 15 rue Victor-Cousin,
75005 Paris, France, in association with the Agency
For Cultural and Technical Co-operation based in Paris,
France. Volumes in the series in the French language
are available from Maisonneuve et Larose.

Published by THE MACMILLAN PRESS LTD
London and Basingstoke
Associated companies and representatives in Accra,
Auckland, Delhi, Dublin, Gaborone, Hamburg, Harare,
Hong Kong, Kuala Lumpur, Lagos, Manzini, Melbourne,
Mexico City, Nairobi, New York, Singapore, Tokyo.

Published in co-operation with the CTA (Technical
Centre for Agricultural and Rural Co-operation),
P.O.B. 380, 6700 AJ Wageningen, The Netherlands.

ISBN 0–333–52313–X

Phototypeset by Intype, London

Printed in China

A catalogue record for this book is available from
the British Library

The opinions expressed in this document and the spellings of
proper names and territorial boundaries contained therein are solely the
responsibility of the author and in no way involve the
official position or the liability of the Technical Centre for
Agricultural and Rural Co-operation.

Front cover photograph of Fulani cow being milked with calf at
foot, courtesy of the author

Sp gravity/ Hydrometer.

TCTA

The Technical Centre for Agricultural and Rural Co-operation (CTA) operates under the Lomé Convention between member States of the European Community and the African, Caribbean and Pacific (ACP) States.

The aim of CTA is to collect, disseminate and facilitate the exchange of information on research, training and innovations in the spheres of agricultural and rural development and extension for the benefit of the ACP States.

Headquarters: De Rietkampen, Galvanistraat 9, Ede, Netherlands
Postal Address: Postbus 380, 6700 AJ Wageningen, Netherlands
Tel.: (31)(0)(8380)–60400
Telex: (44) 30169 CTA NL
Telefax: (31)(0)(8380)–31052

Agency for Cultural and Technical Co-operation (ACCT)

The Agency for Cultural and Technical Co-operation, an intergovernmental organisation set up by the Treaty of Niamey in March 1970, is an association of countries linked by their common usage of the French language, for the purposes of co-operation in the fields of education, culture, science and technology and, more generally, in all matters which contribute to the development of its Member States and to bringing peoples closer together.

The Agency's activities in the fields of scientific and technical co-operation for development are directed primarily towards the preparation, dissemination and exchange of scientific and technical information, drawing up an inventory of and exploiting natural resources, and the socio-economic advancement of young people and rural communities.

Member countries: Belgium, Benin, Burundi, Canada, Central African Republic, Chad, Comoros, Congo, Djibouti, Dominica, France, Gabon, Guinea, Haiti, Côte d'Ivoire, Lebanon, Luxembourg, Mali, Mauritius, Monaco, Niger, Rwanda, Senegal, Seychelles, Togo, Tunisia, Burkina Faso, Vanuatu, Viet Nam, Zaire.

Associated States: Cameroon, Egypt, Guinea Bissau, Laos, Mauritania, Morocco, St Lucia.

Participating governments: New Brunswick, Quebec.

iii

Other titles in *The Tropical Agriculturalist* series

Sheep	ISBN: 0–333–52310–5	Poultry	ISBN: 0–333–52306–7
Pigs	0–333–52308–3	Rabbits	0–333–52311–3
Goats	0–333–52309–1	Draught Animals	0–333–52307–5
Dairying	0–333–52313–X	Ruminant Nutrition	0–333–57073–1
Animal Breeding	0–333–57298–X		

Upland Rice	0–333–44889–8	Sugar Cane	0–333–57075–8
Tea	0–333–54450–1	Maize	0–333–44404–3
Cotton	0–333–47280–2	Plantain Bananas	0–333–44813–8
Weed Control	0–333–54449–8	Coffee Growing	0–333–54451–X
Spice Plants	0–333–57460–5	Food Legumes	0–333–53850–1
Cocoa	0–333–57076–6	Sorghum	0–333–54452–8
The Storage of Food		Cassava	0–333–47395–7
Grains and Seeds	0–333–44827–8		

Other titles published by Macmillan with CTA (*copublished in French by Maisonneuve et Larose*)

Animal Production in the Tropics and Sub Tropics	0–333–53818–8
Coffee: The Plant and the Product	0–333–57296–3
The Tropical Vegetable Garden	0–333–57077–4
Controlling Crop Pests and Diseases	0–333–57216–5
Dryland Farming in Africa	0–333–47654–9
The Yam	0–333–57456–7

The Land and Life Series (*copublished with Terres et Vie*)

African Gardens and Orchards	0–333–49076–2
Vanishing Land and Water	0–333–44597–X
Ways of Water	0–333–57078–2
Agriculture in African Rural Communities	0–333–44595–3

Contents

v

Acknowledgements

I would like to thank the editor of this series, Tony Smith, for his help and comments during the writing of this book.

In addition I am indebted to the following colleagues and friends for their helpful comments and criticisms of drafts of the text: Stewart Macfarlane (Centre for Tropical Veterinary Medicine (CTVM)) for reading numerous drafts of the whole text and Jeroen Dijkman (CTVM), Chris Daborn (CTVM), Denis Fielding (Edinburgh School of Agriculture), Archie Hunter (CTVM), John Maule (Private Consultant), Jamie Bennison (Natural Resources Institute, UK), John Jenkin (Private Consultant) and Richard Clemence (CTVM) for commenting on individual chapters.

Richard W. Matthewman

The author and publishers wish to thank the following who have kindly given permission for the use of copyright material:

Academic Press Inc. (London) Ltd. for Fig 1.1 adapted from *The Biology of Agricultural Systems* by C.R.W. Spedding (1975).

Blackwell Scientific Publications Ltd. for Table 3.5 from *The Calf* by J.H.B. Roy (1980).

Centre for Tropical Veterinary Medicine for a map from 'Mineral Deficiency in Ruminants in Sub-Saharan Africa: A Review' by Schillhorn van Veen & Loeffler, in *Tropical Animal Health and Production*, Vol. 22, pp. 197–205 (1990).

Farming Press Books, Ipswich, for Table 5.4 from *Principles of Dairy Farming* by K. Russell, 8th Edition (1980).

Food and Agriculture Organization of the United Nations for Table 1.3 from *FAO Production Yearbook, 1989.*

W.H. Freeman & Company for Fig. 5.5 from *Biology of Lactation* by G.H. Schmidt (1971).

Longman Group UK Ltd. for Table 6.8 from *Animal Nutrition*, 4th Edition by P. McDonald, R.A. Edwards and J.F.D. Greenhalgh (1987).

J. Oliver for Tables 3.2 and 6.9 from *Dairy Farmers Handbook* published by The National Association of Dairy Farmers of Zimbabwe, Harare.

Photographic acknowledgements

All photographs are courtesy of the author.

Preface

This book is the sixth in a series of livestock books in *The Tropical Agriculturalist* series. This series is intended to provide up to date information for students, extension specialists and farmers. All the books have been written by specialists who have worked in a number of tropical countries or regions. The author of this book, Richard Matthewman, has had considerable experience of ruminant production in both East and West Africa.

This book deals with milk production from cows in various parts of the world. It discusses not only production by large-scale commercial farmers but also by smallholder farmers and pastoralists. It explores the situation of milk production in areas where land is communally owned and competition exists for grazing resources between cattle owners and crop producers. It also covers commercial milk production by both small- and large-scale farmers. Under these latter systems, although land is owned, market and price received for milk affects the economic viability of the enterprise. Planned milk production under these circumstances requires an understanding of the biological processes of milk production, socio-economic and political constraints as well as the simple economic ones.

This book addresses all these issues and when read in conjunction with others in the series, notably the ones on ruminant nutrition, animal health and management of economics of livestock production, the reader will obtain a very good insight into the problems and situation of dairying in tropical countries.

Anthony J. Smith

1 Importance of milk production

Science and practice of milk production

The study of domestic animals can be divided broadly into animal science and animal production. Animal science is the study of the behaviour of domestic animals and physiological functions such as digestion and reproduction, which are similar in milk animals throughout the world, but are influenced by environment and health. Some breeds are better adapted to cooler climates and others to hotter climates. A knowledge of how milk animals function, as well as a knowledge about farmers and farming systems is necessary to plan improvements to animal performance. The study of livestock farmers and farming systems is known as animal production and includes social, economic and political aspects, land-tenure, mixed-farming systems and integrated land-use. This book considers the science of milk production in the context of farmers and farming systems.

> *To plan for improved milk production, advisers and extension staff need to help farmers in ways that will be technically feasible, socially acceptable and economically viable.*

Importance of milk as a food

Milk consists of 80–90% water, but contains important nutrients (minerals, vitamins, fat, proteins and sugars). Composition varies between species (Table 1.1). Those which grow fast usually have higher proportions of protein in the milk.

The energy value of both human's and cow's milk is similar, but amino acid and fatty acid compositions differ considerably. Milk also varies in its mineral (Table 1.2) and vitamin content. Milk is a rich source of these minerals for human diets, but cow's milk has too little calcium which

Table 1.1   Composition of milk from domestic animals and humans

Species	Fat (g/kg)	Proteins (g/kg)	Lactose (g/kg)
Humans	47	13	69
Cattle			
Bos indicus	54	32	46
Bos taurus	44	38	49
Sheep	85	67	47
Goat	45	37	42
Camel	54	39	58
Buffalo	74	38	49

Note   These figures are approximate and conceal the variation between breeds, individual animals and stages of lactation.

Table 1.2   Calcium and phosphorus content of milk

Species	Calcium (mg/100 ml)	Phosphorus (mg/100 ml)	Ca:P (ratio)
Humans	33	15	2.2:1
Cattle	120	95	1.3:1
Goat	130	110	1.2:1
Sheep	200	160	1.3:1

can predispose to low blood calcium levels (hypocalcaemia) in infants fed solely on cow's milk.

Despite differences in the composition of milk from domestic animals and milk produced by humans, the former provides an excellent food for human beings. Milk complements all diets and is valuable for growing children, convalescing adults, pregnant and lactating women and for older people. It can be used in many recipes and made into milk products such as cheese, yoghurt and ice-cream. For these reasons the value of milk as a human food cannot be over-emphasised. However, there is evidence that some people lack the enzyme lactase, which breaks down lactose (milk sugar), and they have difficulty digesting milk.

Importance of milk production in tropical regions

Only 20% milk production from animals for human consumption occurs in tropical regions, even though these parts of the world have 74% of the human population and 69% of the bovine populations (Table 1.3). Efforts are being made in a number of countries to expand home production, with the result that the current rate of increase of milk production is

Table 1.3 Estimated world milk production for consumption by humans

Region of the world	Milk production (% of world total)
North America	17
Western Europe	30
Eastern Europe and former USSR	33
Africa	2
Latin America	9
Near East	2
Far East	7

Source: FAO (1988)

greater in tropical countries than in the traditional milk producing and exporting nations of the world. The number of countries whose governments have implemented effective dairy development policies has been greater in Asia than in other continents.

In the early 1980s, the European Community (EC) and New Zealand each contributed 33% of the total world milk exports although together they only produced 26% of the total world milk. Surpluses of the EC have been reduced as a result of the introduction of milk quotas.

The share of livestock production in the agricultural Gross Domestic Product (GDP) of sub-Saharan Africa in 1980 was estimated to be 17%. The expansion of milk output in sub-Saharan Africa has not kept pace with the increase in human populations, the rates of increase of which are 1.4% and 2.9% respectively.

The shortfall in production has been met from imports of milk products, which increased sixfold in the decade prior to 1984. Sub-Saharan Africa imports roughly 30% of its total milk for human consumption. Disincentives to domestic milk production occur where competition exists between imported and locally produced milk. Farmers must receive a fair price for milk if their enterprises are to be viable.

In addition, price differential increases between meat and milk leads to milk becoming less favoured than meat for home production. This discrimination has been encouraged by better transport, marketing and availability of alternative food products in remote areas. The need for self-sufficiency has been questioned in some countries and this may remove the incentive for farmers to invest in dairy enterprises.

Milk production systems

Milk production is influenced by cattle distribution, which in Africa has depended on the distribution of tsetse flies and sleeping sickness (trypanosomiasis). Tsetse flies are found in wetter, lowland areas and so cattle have been restricted to dry areas such as the Sahel, parts of East and Southern Africa and highland regions in Ethiopia, Kenya, Tanzania and Malawi. Historically, milk production has also been restricted to these areas.

Milk is produced in **extensive, semi-intensive** or **fully intensive** systems. Extensive systems include pastoralism in drier regions of Pakistan, Africa and the Middle East, village milk production in which cattle are herded around the village on communally owned land and dairy ranching systems in South America in which 'dual-purpose' cows are milked. Semi-intensive, smallholder production is where farmers may own only one or two cows and cut-and-carry food to their animals. Intensive systems utilise high levels of inputs and management skill. Milk is produced from cattle, buffaloes, sheep, goats and camels depending on the country and system.

Pastoralism is an important way of life in dry places, where cattle owners (pastoralists) are often transhumant. Milk contributes to subsistence food supplies and provides food in areas where other foods are often not available. Milk is produced and can be traded for foods such as sorghum (Guinea corn), millet, pulses and vegetables.

In contrast to this, milk has played only a small part in the diets of people in some wetter areas where tsetse flies are found and which have largely excluded cattle. Large ruminant livestock have made a smaller contribution to the generation of wealth in such areas and animal protein supplements are derived from poultry, pigs, small ruminants, fish and game (bush-meat).

In settled smallholder systems, in tsetse-free areas, milk animals are important. Milk is used for daily home consumption and cash income. Milk animals help to utilise spare labour in offpeak seasons and increase the efficiency of use of land, crop residues, by-products and fodder from fallow land. Cattle dung contributes to soil fertility. At higher levels of output per animal, milk competes favourably with other forms of livestock production in terms of efficiency of nutrient conversion (Fig 1.1).

Limits to increased milk production

The factors which limit milk production are known as **constraints**. These include environmental, social, economic, marketing, technical, husbandry, management, biological and genetic constraints.

Traditional systems are controlled by natural conditions, such as availability of food and water, climate and disease, to which cattle must be well adapted. The level of milk production is usually low because of the

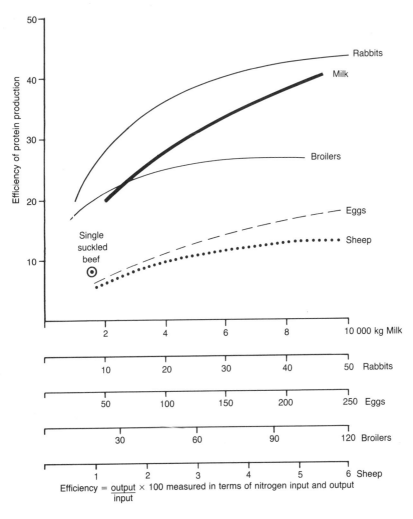

Fig 1.1 *Efficiency of milk production compared with other forms of production in different species of domestic animals*
Adapted from Spedding (1975)

quality of food inputs, environmental stress and the low genetic potential of cattle to produce milk.

To increase milk production, it is necessary to remove limiting factors.

> *The degree to which constraints can be overcome depends on the farmer's knowledge and ability and the capital available for investment.*

Constraints such as high ambient temperature and relative humidity can be reduced by providing housing or shade. Constraints such as poor nutrition, disease risk and infertility can be reduced by intervention, which requires knowledge and often some financial input by the farmer. Management constraints can be overcome by extension work or by encouraging farmers to attend training courses to increase knowledge and ability. Once constraints are overcome, there may be a case for improving the genetic capacity of stock.

Overcoming constraints is one of the underlying principles of livestock development, but many attempts to increase production have failed. Past attempts should be seen as 'lessons well learnt' and should not deter interested farmers from investing in milk production. With the experience of past schemes and adequate economic incentives, milk production is within the grasp of many farmers.

Objectives of this book

In the following chapters, the principles of milk production are outlined and the problems which face farmers and the way that these may be overcome are discussed. Detailed methods of production are not covered because of the diversity of production systems. Milk production by sheep, goats, buffaloes and camels is not discussed because this topic is covered by other books in this series (*The Tropical Agriculturist*).

The term milk production is used rather than dairy production, since the latter term applies more to intensive methods of production and the use of dairy breeds which have been bred for their superior milk production.

Outlines are given of examples of milk production systems in Chapter 2. These 'case studies' indicate the diversity of methods of milk production and act as a reference point for later chapters.

Constraints.
* genetics cows often not breed for milk prod^n.
* food subs not 4 quality.
* Env^m not 6 stress.

2 Systems of milk production and marketing

Milk production systems are grouped under the headings of pastoralism, dual-purpose, dairy ranching, smallholder and intensive production. Milk collection schemes are described as ways of increasing offtake and of stimulating increased production. How milk is produced and marketed under different climatic conditions where different social and economic constraints prevail are described for 16 examples in Chapter 2.

Milk production in pastoral systems

In dry areas where crops cannot provide fully for **subsistence** needs or cannot be grown at all, milk from cattle (and sheep, goats and camels) also contributes to subsistence needs. Livestock production is often the only way to use dry areas and milking animals allows owners to survive in dry areas. High yielding cows are unsuited to such environments and natural selection has favoured hardiness and adaptation rather than high levels of production.

> *Pastoralism is a milk production system in which well adapted indigenous cattle are herded in arid and semi-arid areas by traditional cattle owners who depend on milk to meet part of their subsistence food requirements. The productivity of the cattle depends largely on natural, environmental factors.*

Pastoralists traditionally have been **nomadic** or **migratory** to take advantage of seasonal and regional variations in grass growth, rainfall and water availability. They are becoming increasingly **transhumant** or **semi-settled** to lay claim to land in higher rainfall areas. Milk production is seasonal and yields are low. The ability of their animals to produce some milk is more important than high yields. Since yields are low, there must be a balance between milk taken for human consumption and by the calf. Calf growth is usually low and mortality high.

Animal populations fluctuate because of unpredictable climate and periodic disease outbreaks. As a result, pastoralists increase stock numbers as a safeguard against risk. Large numbers are also a sign of wealth. Each family unit requires enough cows to provide milk and to ensure herd replacements.

The characteristics of livestock systems which have evolved in dry areas make 'planned development' difficult. The systems works well when population densities are low. Few examples exist in which 'planned development' has resulted in reduced risk and so pastoralists continue to operate according to traditional methods. There are good arguments why pastoral areas should continue to be managed in traditional ways by the people who know the conditions best and in whose interest it is to maintain the systems. Competition for land with crop producers in wetter areas, high human and stock densities in drier areas and communal land use make land management difficult.

Examples of milk production in pastoral systems

The complexity of pastoral systems and the problems which have arisen during development efforts are illustrated by a consideration of systems in different countries. The following case studies of Nigeria, Uganda and Kenya highlight the characteristics. Since land is *communally* grazed there is no incentive to improve land. In dry areas where crops cannot be grown, overgrazing and removal of woody vegetation for firewood has led to increasing soil erosion. Even though milk is the basis of pastoral economies, increasing production often presents insurmountable difficulties. Maintenance of the present *status quo* or return to a previous *status quo* might be the most feasible objective for planners.

Nigeria, West Africa – Fulani people

Seasonal changes in rainfall, pasture growth and animal diseases have determined the pastoralists' way of life for the Fulani in Nigeria. The Fulani have traditionally moved south in the dry season and north in the wet season, to find grazing and water and to avoid the presence of tsetse flies (*Glossina morsitans*, *G. palpalis* and *G. tachinoides*) in southern areas of the middle belt. Fulani herds can still be seen moving north in July in Kano State to their wet season grazing lands. The average one-way distance moved by Fulani in Sokoto State is reported to be 120 km (minimum 10 km, maximum 300 km). Fulani with small herds have often grown some crops (mainly bulrush millet) to supplement grain procured by bartering milk. The growing of crops requires agreement with the local settled crop farmers.

From 1960 to 1990 human populations in Nigeria have expanded greatly: this has resulted in more land being used for cultivation and less land for grazing. Population changes have contributed to the reduction of tsetse flies by destroying the vegetation and habitat (required by tsetse flies) and killing game (hosts of tsetse flies). This has allowed the Fulani to remain longer each year in more southerly areas. In central and southern Nigeria (Kaduna, Niger and Oyo States) the Fulani are becoming settled, mixed farmers. Fulani people are now found in many areas south of the River Benue and River Niger.

Cattle are milked at dawn. Milk is used sour and in combination with other foods. Women sell milk in markets or barter it for staple grains and other foods. A family requires a minimum of 45 cows, of which half need to produce milk in any one year. The number of cows required may

Fig 2.1 *Fulani cattle owner in Anambra State, Nigeria*

9

Fig 2.2 *Fulani women with milk in calabashes*

be higher than this, particularly in drier years when cows produce less milk. Less milk is produced in the dry season than in the wet season and families rely more in the dry season on grain bartered earlier in the year for milk.

The cattle are grazed for varying periods each day depending on season. The longest grazing time occurs in the early wet season when the grazing time can be up to 11 hours a day and the shortest time, about seven hours, occurs in the late wet season. Walking accounts for about 20% of the herding time, grazing 75% and resting and watering 5%.

There is no deliberate breeding season and bulls have access to cows all year round. Usually there is more than one bull in a herd and males are only castrated when they become difficult to handle. Cows may calve in the *ruga* (temporary settlement) or when out grazing, but after calving remain in the *ruga* for 24 hours or more so that the calf can suckle regularly. The cow is fed grass or tree leaves during this period. Thereafter the calves remain in the *ruga* during the day for three or four weeks and are then allowed to accompany the cows to graze.

Fulani are good animal husbandry people and their expertise is a good foundation for the development of mixed smallholder cattle/crop systems. The Fulani in Nigeria illustrate the transition from nomadic pastoralism to more settled systems of milk production. Within the constraints of the present land-tenure and agricultural systems, there is little chance that the traditional pastoral system can survive in Nigeria, except in drier and more northerly areas. Settled farmers have greater land rights than pastoralists who must settle and integrate into the established land-tenure system to gain rights to land use. Examples of this exist in Niger and Oyo States, where Fulani have been settled for a number of generations. Settled crop farmers often own cattle which are tended by the Fulani. In traditional pastoral areas, a ready market usually exists for milk, since the local people have become accustomed to milk as a food. Where pastoralists have not been present, such as in the more southerly Eastern States, milk is not used by the local people and a ready market does not exist. Here Fulani women have been observed to keep larger numbers of chickens as a source of cash income.

Draught animal power is becoming more important in Nigeria and White Fulani (Bunaji) cattle are well suited for this purpose. The prospects for milk production in Nigeria and the improvement of the cattle industry are good, because rainfall in the south is high, mixed farming is increasing and there is greater potential for food production for milk cows than in drier areas. There is also the potential for the development of milk collection schemes (p. 28) though this has been attempted in the past without success.

Uganda, East Africa – Karamoja region

Pastoralism in Karamoja illustrates the problems associated with the 'development' of pastoral systems. Karamoja occupies 35 000 km² in north east Uganda. Annual rainfall is high in the west (over 1000 mm) to less than 500 mm in the east, with the dry season between October and March. The people are mostly transhumant, semi-pastoralists: their main settlements are in the central belt of Karamoja where water is available most of the year. This belt acts as a *cordon sanitaire* between the groups: the women, elders and children live there all year round; and the young male herders only live there during the wet season. Cultivation of sorghum, vegetables, groundnuts, maize and millet occurs along the rivers. Erratic rainfall makes it impossible to rely solely on crop production for subsistence as droughts occur one year in four. An unstable balance exists between attempts to increase herd size (for milk, draught and to safeguard against drought) and losses from disease and droughts.

In the early part of the twentieth century (if the historical catalogue

11

of events is to be believed) a series of 'development catastrophies' occurred. The government of the day wished to increase productivity and make the people more accountable to government. The implemented policies failed to consider the fragility of the pastoral system. The result was that the pastoralist way of life was disrupted and the resource-base was denuded. Periodic droughts continued, human and cattle populations grew and the situation deteriorated further. Further political strife (late twentieth century) added to the problems faced by the people of Karamoja. The failure of the attempts to bring about 'planned development' of the pastoral systems in Uganda provide lessons for livestock planners.

> *The balance of pastoral systems cannot be quickly changed or manipulated, if at all. It is unlikely that static production systems in dry areas can be sustained. Only flexible systems, which can utilise the resources of large areas will be able to survive.*

Kenya, East Africa – Maasai people

A number of pastoral systems occur in Kenya, based on different ethnic groups, but the problems facing the people are much the same as those facing other pastoralists in Nigeria and Uganda. The government has tried to influence the management of pastoral systems by establishing group ranches as part of the policy of privatisation of land ownership. In Kajiado and Narok, ranches were established in the mid-1960s and consisted of groups of Maasai families with approximately 200 ha/family. The Maasai have traditionally occupied a large area in Kenya and Tanzania, but since the 1960s the area has been reduced by crop production and restrictions imposed by the positions of game reserves. These units of 200 ha have proved to be too small and do not take into account periodic droughts. When drought occurs, the Maasai have to move to areas where conditions might be better, but this is not possible with a settled ranch system.

Despite considerable development input over a decade or more and efforts aimed at commercialisation, the production strategies of the Maasai are still geared to supplying subsistence needs with a relatively low level of market offtake.

> *Unpredictable climate, communal grazing patterns and pressure on land for crop production restrict flexibility of pastoral movements, yet force pastoralists to move.*

Kenya – Ngisonyoka Turkana people

A different system from that of the Maasai exists in northern Kenya, where the Ngisonyoka people, a sub-section of the Turkana tribe, are found. In 1982 the tribe consisted of approximately 9650 pastoralists, 85 200 sheep and goats, 9800 cattle, 9800 camels and 5300 donkeys on 7540 km² land. Families move up to 15 times each year and may cover over 100 km. Mean annual temperature is 30°C and rainfall 150–600 mm depending on location. Milk supplies 60% of food energy for humans, of which half is from camels, a quarter from sheep and goats and a quarter from cattle.

The following conclusions have been drawn about the **energy balances** and **efficiency of energy transfer** through the system.

- Pastoralists utilise the area's resources so effectively to permit a high density of humans on marginal land without causing soil degradation.
- No causal apparent relationship exists in this case between pastoralism and environmental degradation.
- Herd sizes are not excessively large.
- Non-lactating animals are not in excess.
- Low animal productivities do not result in a high ratio of livestock to humans; the observed value is 3.8 tropical livestock units (TLU = theoretical animal of 250 kg liveweight) per person, but this can still provide enough energy to sustain human populations.

Whether these conclusions are valid for all pastoral systems is doubtful, but if pastoralist groupings are optimum, pastoralism offers a good management system for arid areas. If human and stock densities are not too high, balanced systems can exist and continue to produce livestock products for subsistence and barter.

Kenya – Borana people

A further pastoral system exists in Kenya in the Isiolo District where the Borana people predominate. A well organised system of land and communal stock management operates successfully, though there is competition for land from other pastoralist groups (e.g. Somali). People and land are divided into community groups, *Dedas*, and cattle are herded communally within each *Deda*. Dry herds are herded separately from milking herds, the latter staying nearer to the community base where the family groups are found. Men herd dry cattle further away to take advantage of grazing and water resources. The area is too dry for crops except along rivers where irrigation is possible and so milk is important for the livelihood of these people. Local government boundaries have been drawn to maintain the integrity of the pastoralist groups to allow communal management of land resources without interference from other groups. This system offers a model for the management of similar arid areas.

Summary of characteristics and pastoralism

The common characteristics of the areas where pastoralism is practised are listed below:

- Past planning has assumed that pastoral systems could or can be 'developed' or 'changed for the better'; this in most cases has been shown to be untrue and that the traditional systems are very durable at low population and stocking levels.
- Human and stock populations have increased and land is under pressure from both pastoralists and crop producers.
- Land degradation has occurred in many pastoral areas, because of overstocking linked to poor land-use planning, climatic change and removal of woody vegetation for firewood.
- Transhumant movements are decreasing in distance and frequency and pastoralists are settling and becoming more dependent on crop production.
- Traditionally, when population pressure was lower, pastoral systems probably converted the available sunlight, rain and soil minerals as efficiently as was possible into human food in the semi-arid areas concerned.
- Traditionally managed pastoral systems probably operate optimally and few methods exist to significantly increase the primary or secondary productivity.
- A return to traditional control mechanisms would be appropriate. This realistically means that pastoralists should be allowed to manage their areas according to traditional methods, which implies reduced human populations, reduced stock numbers and controlled stocking rates. Good husbandry and reduced mortality from diseases would help to reduce risk and increase confidence, so that pastoralists could plan their enterprises to meet their needs.
- More effective methods of soil conservation, methods of reducing stock numbers, increased offtake and better control of land-use for livestock could be envisaged if risk could be reduced and the cooperation of the pastoralists could be gained.
- There may be some merit in providing buffers for pastoralists against uncertainty (ready markets in times of stress, insurance against drought or epidemics, provision of food supplements for stock), but these suggestions would be difficult to implement in many areas.
- Incentives for pastoralists to turn to other means of employment would be appropriate, but would be impractical in countries where job opportunities are scarce.
- Milk from cattle, camels, goats and sheep remains an essential product of pastoral systems. Any future strategy for pastoralists must consider

ways of providing an adequate income from milk or alternative sources of income for the family unit. The number of cows required to sustain a family in a traditional system can be as high as 70.

- Livestock planners would be wise to consider development attempts that have been made in other pastoral systems before embarking on ambitious development plans.
- Comparisons with sustainable cattle ranching systems in dry zones of countries, such as Australia, are valid in terms of goals for technical production aspects, but not in terms of social aspects and human welfare in pastoral areas.

Dual-purpose systems

In the pastoral systems (pp. 7–14) milk provides the livelihood of the pastoralists and sustains the system. In dual-purpose systems, the importance of milk is shared with other products. Two systems are described – dairy ranching and milk from draught cows. In the former, the enterprise produces milk and beef and in the latter, milk cows are used to produce draught power for cultivation or transport.

Dairy ranching

Dairy ranching is an extensive system where land is owned, but where conditions do not favour intensive milk production. Animals are milked once a day and the calf is reared by the cow in a similar way to that described for pastoralism. The system is intermediate between extensive beef production and intensive dairying with specialised dairy breeds and twice a day milking. It is a dual-purpose system utilising local and crossbred cows in climatic areas where high yielding exotic breeds are not appropriate and where management is not adequate.

Dairy ranching is a method of milk production in which low to moderately yielding Zebu or Zebu-cross cows, maintained in extensive beef ranching systems, are milked once a day and they each rear a calf.

The method is practised widely in tropical areas of South America where milk is produced from low to moderate yielding cows in dual-purpose beef–milk systems. Approximately 3 million km² of non-arable tropical savannas are used for stock production. In these regions, which lie mainly below an altitude of 2000 m, extensive beef ranching occurs using Zebu breeds and their Zebu-Criollo crosses. In Nicaragua 70–80% of lactating cows are milked, while in Colombia over 50% of the milk consumed comes from dual-purpose systems. In Brazil the corresponding figure is 35%. European breeds are not commonly used in the lowland regions, but in highland regions (exceeding 2000 m) conditions permit the use of European breeds for milk production in intensive systems.

Depending on the market for milk and the food supply, a variable number of cows in the beef herd may be milked once a day in the rainy season. The calves are allowed restricted access to the cow during the daytime. The system flourishes best where both land and labour are cheap and plentiful and a good price is paid for milk. The prospects for increased milk production from dairy ranching appear to be promising and the methods are relevant to other parts of the world.

Examples of milk production in dairy ranching systems

Colombia, South America

Colombia is a country of widely differing topographical conditions with regions which vary from sea level with hot climates to regions between 2600 m and 3000 m above sea level with cool climates and regions with snow at altitudes above 3000 m. Out of 27 million cattle in Colombia, it is estimated that there are 3.7 million dairy cattle with 2.5 million in milk production. Approximately 9% are select dairy breeds, 58% cross-breds and 33% the Criollo (local) breed. Only 60% of milk comes from dairy herds, the rest from beef cattle. In cool climates (valleys and plateaux, 1500–2800 m) the milk animals are predominantly Holstein, Ayrshire and Brown Swiss breeds, whereas in hotter areas (hot valleys, the north coast and foothill regions) milk animals are mainly dual-purpose milk and beef Criollo, Zebu and crossbred animals. In some coastal areas pure beef breeds are milked. The system varies with the region and depends mainly on climate and distance from markets. In the Caribbean and Piedmonte Regions production is mainly extensive, whereas in the Andean Region more intensive farming is carried out. Year-round calving occurs and income can be generated from milk throughout the year. Up to a third of farm income can be derived from milk.

Bolivia, South America

In the humid sub-tropical area of San Javier, Santa Cruz Department of Bolivia 500–600 m above sea level of latitude 16°S, dairy ranching comprises Zebu/Criollo cows milked once a day at daybreak with the calf at foot. The calves run with the cows during the day, but are separated at night. Average milk sales are low being approximately 400 kg during a 180-day lactation.

Fig 2.3 *Dual-purpose Criollo cattle in a dairy ranching system, in Bolivia*

The maintenance of Holstein/Friesian cattle in the Bolivian lowlands is not profitable, but crossbreeding Zebu/Criollo cows with European bulls results in the first two generations in vigorous crossbreds that can produce milk economically. With good management such animals produce more milk than purebreds, tend to be more fertile and have significantly lower calf and adult mortality. Crossbreds are cheaper to maintain than purebreds, produce similar yields at a lower cost and do not have the high risk involved with importing exotic cattle into the tropics. Dairy ranching is therefore ideally suited to the lowland tropical areas of Bolivia.

Summary of characteristics of dairy ranching

- Dairy ranching is an adaptive method of milk production. It exploits the milk producing qualities of better cows in beef herds of adapted breeds in areas of lower potential with higher temperatures.
- Both good markets for milk and transport systems are necessary. In South America, milk is often transported long distances by rail to large urban centres.
- Where dairy ranching occurs in South America, land is owned and farmers have responsibility for the upkeep of land.
- Milk can represent up to a third of farm income. Such income is more regular than that from sales of cattle for beef.
- Better cows may produce 800 litres of milk for sale in a lactation without calf growth being adversely affected.
- Once a day milking is carried out and the calf is reared by the cow.

17

Milk production from draught cows

Draught animals are used in many parts of the world to produce power for cultivation and transport. Usually bulls (or oxen) are the preferred draught animal. Where land for agriculture is scarce, farmers are forced to use cows – this allows the food available for livestock to be used more efficiently to support cows rather than both cows and bulls (or oxen).

Bangladesh, Indian sub-continent

It is estimated that more than half the draught animals are cows in Bangladesh. High human population pressure on land in this country has reduced average farm size to less than 2 ha and cropping intensity is high, with two or more crops planted each year. Small farms do not have the capacity to produce enough fodder for both cows and bulls. Farmers have to sell male animals quickly and cows are used for work in addition to producing calves and milk. This increases the energy requirements of cows and often means that such cows do not receive enough food to meet their full energy requirements. In Bangladesh, draught cows are usually small and undernourished. As a result, milk yield can suffer, as well as fertility.

Such systems deserve concerted attention from farmers to find ways of achieving sustainable levels of production, because of the higher demands placed on draught cows.

Smallholder systems of milk production

Adapted local breeds of milk cows are kept throughout the tropics by settled crop farmers on small farms or in villages or peri-urban areas. Sometimes crosses with Jersey, Sahiwal or other higher yielding breeds may occur. The cows may be tethered on the farm, herded on communal grazing land or stall-fed in the village or compound. Farmers usually own a few cows to complement crop farming. They use the milk and derive income from sale of milk, male calves or fattened steers.

> *Smallholder or backyard milk production is characteristic of farmers who keep small numbers of cows herded near the farm or fed cut-and-carry grass and crop by-products and milked for family use or local sale.*

Smallholder production can begin to form the basis of a commercial dairy industry and a number of smallholder milk schemes have been initiated with varying degrees of success. Such schemes require support services from government and commercial enterprises for food and drug supplies and milk processing and marketing.

Examples of milk production in smallholder systems

Kenya, East Africa

Prior to 1965 most milk in Kenya came from large farms based on extensive and semi-intensive grazing. In the 1960s and 1970s Kenya exported milk products to neighbouring countries, but this declined in the mid-1970s. Some 80% of milk (early 1990s) originates from small-scale mixed-farms. The degree of specialisation is influenced by the profitability of milk production compared with other crops, such as tea, coffee, pyrethrum, maize and beans.

Agricultural land in Kenya is classified into high, medium and low potential land. High potential lands receive over 850 mm rainfall per annum and occupy 13% of the agricultural land. Smallholders (average 1.2 ha) in high potential areas account for 70% of the population, 75% of agricultural output and 70% of employment.

Levels of specialisation include subsistence, smallholder and commercial milk production. Smallholder grazing systems in the highlands are based on alternation between arable cropping and fallow. A thriving industry has developed based upon grade cattle (usually Local × Friesian). These are not well adapted to lowland and marginal areas, where *Bos indicus / Bos taurus* crosses are recommended. The Sahiwal has been introduced and Local × Sahiwal cattle are used for milk in coastal plains where farmers grow Napier grass (*Pennisetum purpureum*) and leucaena (*Leucaena leucocephala*) for their cows. The Ayrshire × Sahiwal is used and other crosses involving Sahiwal, Jersey and Brown Swiss have been used successfully.

Table 2.1 Composition of Kenya national dairy herd

Breed	Numbers
Ayrshire	400 000
Friesian	337 000
Guernsey	224 000
Jersey	117 000
Exotic × Zebu	410 000
Total	1 488 000

In Western Kenya, cattle are kept by the Abaluya and Luo peoples, but herd sizes are declining because of high human population pressure on land, particularly in the Siaya and Kakamega Districts. Large-scale sugar production has also excluded cattle in those areas.

19

Fig 2.4   *Ayrshire cattle, in Kenya*

Approximately 60% of milk is consumed on the farm and only 20% finds its way to the formal market, the remaining 20% being sold unprocessed locally. The Kenya Co-operative Creameries pay lower prices than other consumers and have tended to be a buyer of last resort for farmers after local demand has been satisfied.

Grazing systems based on sown leys or indigenous pastures have given way in suitable areas to stall-feeding and zero-grazing. About 80% of smallholder cattle are stall-fed for most of the year. Their diet includes Napier grass and other fodders including maize stover, banana stems, sweet potato vines (*Ipomoea*) and other arable by-products.

Large units have natural pastures of Kikuya grass (*Pennisetum clandestinum*) and star grass (*Cynodon nlemfluensis*) and couch grass (*C. dactylon*) in suitable areas and planted leys of Rhodes grass (*Chloris gayana*), nandi setaria (*Setaria anceps*) and Guinea grass (*Panicum* spp.) elsewhere. Milk production is based on grass for seven to eight months and then relies on conserved forages (silage and hay) or cut fodders (maize and Napier grass) during the dry season.

Pure exotic stock and high quality crossbed stock (those produced by sequentially mating indigenous stock to purebred bulls) are kept in areas of high potential. Artificial insemination (AI) is practised and 700 000 doses of semen are distributed annually (1984).

Malawi, South East Africa

In Malawi prior to 1975 large amounts of dairy products were imported and this stimulated the establishment of a dairy development programme in 1971 by the Food and Agriculture Organisation of the United Nations

20

(FAO) in conjunction with several local government departments. This involved the establishment of government dairy farms and the promotion of smallholder production by the release to farmers of Friesian × Zebu first-cross cattle. This scheme involved crossing Malawi Zebu cows with Friesian bulls on government farms and inseminating the resulting first-cross heifers with Friesian semen. Four weeks after calving those cows which passed a minimum standard of milk production were sold to selected farmers or funded using repayable loans.

Before receiving a crossbred cow the farmers had to have adequate finance and participated in a dairy training course; the farm had to be situated not more than 8 km from the milk collection point, have a good water supply and have a minimum of 3 acres (1.2 ha) of Napier grass/Rhodes grass/leucaena under cultivation.

The farmer also received a hand-spray gun and 2 litres of acaricide, so that cows could be hand-sprayed weekly. The cows were zero-grazed with cut-and-carry grass/browse and supplemented with crop by-products (e.g. groundnut tops, maize bran or maize stover) and compound dairy concentrate if available. Farmers were encouraged to make silage or hay. The AI service was provided free of charge.

In April 1988 there were 356 smallholders with crossbred dairy animals in the Central Region with a total of 642 adult cows. The farmers were organised into geographically convenient groups, each run by an elected committee of farmers and had a milk cooling centre. Cows were milked twice daily and the milk taken right away to the nearby cooling centre from where a Malawi Dairy Industries tanker collected milk each second day. Farmers were paid a set price per litre with a bonus for above average fat. Loans taken out for the purchase of cows were repaid by deduction of a sum from the monthly milk cheque to the farmer, usually over three years. Milk yields recorded over a ten year period (1973–83) from a total of 781 lactations recorded, showed the mean milk yield of crossbred cows to be 2188 kg in a mean lactation of 392 days, with mean daily yield 5.7 kg.

Although adequate provision for providing food for cows was made initially, many farmers increased their cow numbers, but could not increase land and pasture holdings. These farmers found it difficult to maintain crossbred cattle and the scheme after a few years was found to be losing momentum.

Summary of characteristics of smallholder systems

- There is considerable potential for diversification and increased production in small-scale settled systems. This may involve the use of planted feeds, crossbreeding, artificial insemination and improved milk marketing and hygiene.

- Where cows are kept on smallholdings as a complement to settled crop production, there is potential for support services to increase levels of production.
- Cows produce low levels of milk, but there is potential for improvement if ways can be devised of planting forages on small plots around the smallholding.
- Smallholder milk schemes have been popular in many countries, but have not always been as successful as anticipated. Success depends on good planning and technical and economic feasibility. Many examples exist on which to base future plans for smallholder dairy schemes and from which much can be learnt.
- Criteria for success include the provision of a guaranteed price for milk for the producer, the lack of competition for the use of land between crop production and food production for milk cows, an adequate infrastructure to allow the transport and central collection of milk, a good extension and back-up advisory service from the Ministry of Agriculture and an assured supply of necessary inputs such as drugs, food supplements, pasture seeds and equipment.

Intensive systems of milk production

In extensive and semi-intensive systems (described earlier) milk production per cow and per farm is often low, but is balanced by low levels of inputs and managerial skill. In semi-intensive systems some inputs are used in greater quantities than in more extensive systems. In fully intensive systems all inputs are used in greater quantities than in extensive and semi-intensive systems – animals receive more food, better health care, better housing and greater attention.

> *Intensive systems use high levels of inputs and require high levels of managerial skill to overcome constraints and to raise the level of productivity to cover additional costs of the enterprise.*

Once investment increases, production and profits must also increase. In intensive systems, management control is high and to achieve high levels of production, requires skill and dedication. Irrespective of their genetic potential, cows must receive enough food for their requirements and be well managed. The age at first-calving must be less than 3.5 years, calving intervals less than 450 days, mean lactations longer than 200 days and female mortality to first-calving less than 10%. Such systems might also aim for a year round supply of milk to offset year round costs of production.

Examples of milk production from intensive systems

Nigeria, West Africa – Vom region

Vom is situated on the Jos Plateau in northern Nigeria and is cooler than the surrounding Guinea savanna plains. Tropical *highland* areas are cooler and hence advantageous for livestock which suffer less environmental stress. Conditions are also better for producing and marketing milk.

Investigations into the relative merits of different breeds commenced at the Livestock Improvement Centre at Vom in 1925 and later attention focussed on the White Fulani which is predominant in northern Nigeria. A yield of 450 kg/lactation and later 570 kg/lactation was set as a minimum standard for cows in the herd. Improvement on the basis of selection was slow and in 1952 up-grading commenced using Friesian bulls and artificial insemination. Until 1970 Friesian bulls were imported regularly and acclimatised readily.

Cattle were grazed on sown pastures in paddocks varying in size from 1.2 ha to 10.0 ha. Most were under permanent grass comprising *Andropogon gayanus, Cynodon dactylon, Chloris gayana, Hyparrhenia rufa, Panicum maximum, Pennisetum clandestinum, Pennisetum purpureum, Stylosanthes guianensis* (see pp. 86–88). Some 500–600 tons silage (mainly maize, but also grass and legume) and large quantities of hay were produced annually up until 1970 for between 220 and 320 head of cattle. Acha straw (*Digitaria exilis*) was also bought from local farmers. A production ration of concentrate (based on maize, Guinea corn (*Sorghum vulgare*), groundnut cake and cottonseed) was fed when required and mineral supplements were fed year-round. Calves were bucket-fed from three days old. Milk let-down problems in White Fulani grade cattle at Vom were never a problem. Halfbred Friesian cows had an average milk yield of 1700 kg/lactation compared with 780 kg for White Fulani cows. The fertility and health of crossbred cows were generally good.

Most of the milk produced was supplied to the Nigerian Creameries Ltd (Madara Ltd), where it was pasteurised and cartoned for the market in Jos city. Originally a number of collection centres for cream and butterfat were established 50–80 km around Vom. The number was reduced in 1961 and later the creamery ceased to function due to marketing problems and poor management. A new Livestock Investigation and Breeding Centre was established in 1980. The original farm ceased to function because of management problems and the high cost of inputs and imported machinery. Much of the original pasture became cultivated by local farmers. The early dairy successes at Vom, however, demonstrate the technical feasibility of milk production in the area.

Malaysia, South East Asia

Milk production in Malaysia is interesting because it is an example of a

23

newly developing industry. Traditionally fresh milk has not been consumed by the Malay and Chinese communities and most of Malaysia's milk has been produced and consumed by the Indian community. These farmers keep Local Indian Dairy cattle (LID) which are distinct from the native Kedah/Kelantan cattle.

In 1969 milk production was estimated to be 4 million gall/year and the requirement to be 60 million gall/year (18 million and 273 million litres/year respectively). Much of the short-fall was met with imports and by 1980 it was estimated that imports would be 80 million gall/year (363.7 litres/year). It was considered that fresh milk production and consumption could be increased if quality could be increased and prices decreased. It was intended to build a dairy industry using intensively managed crossbred animals.

In 1986 there were 36 Milk Collection Centres in the country modelled on the system used in Anand, India (see p. 29). The first centre was at Jasin near to Malacca town. The objective of the dairy development programme was both income generation for farmers and milk production based on improved methods developed from the traditional Indian systems. In 1980 there were approximately 5000 small-scale dairy farmers each with an average of 5 crossbred Friesian/Local or Sahiwal/Friesian cows.

In Kelantan State small-scale units have been established using LID × Friesian cattle. Fodder crops Napier grass (*Pennisetum purpureum*) and Columbus grass (*Sorghum almum*) are grown for green feeding and for silage.

Fig 2.5 *Cross Friesian cattle, in Malaysia*

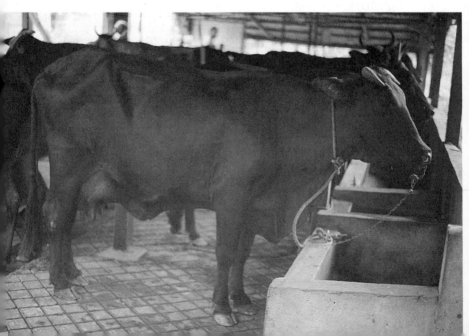

On similar units (area 1 ha) in Serdang, Selangor State, cows grazed *Setaria sphacelata* or *Brachiaria decumbens/Leucaena leucocephala* or were fed cut-and-carry fodder. First-lactation milk yields were up to 1800 kg depending on feeding regime; older animals fed concentrate supplements produced up to 3350 kg/lactation. With high levels of bought-in food, milk yields measured on a per hectare basis were as high as 16 000 kg/year.

Breed improvement is an important component of intensive systems and in 1972–73 Australian Milking Zebu (AMZ) cattle were imported from Australia with a view to upgrading cattle. Not many animals have been distributed since farmers prefer LID/Friesian cross animals. The reasons are not clear, but may relate to the superior beef producing quality of male calves from the LID × Friesian.

Mozambique, South East Africa

Intensive dairy production has been carried out at the University of Macaneta Station, near Maputo (latitude approximately 26 °S). Mean monthly temperatures at the Station vary from 21°C in June–July to 28°C in February–March. In summer temperatures reach 35°C and occasionally 40°C and higher. Annual rainfall is approximately 760 mm and relative humidities are high. In the dry season (April–September) precipitation is negligible. These conditions are not favourable for milk production and management must be good to achieve reasonable productivity. The cattle kept were Friesians.

Cows were rotationally grazed all year on paddocks near the farm. This avoided excessive walking by the animals and so reduced heat stress. Shade was provided by trees or thatched sheds and drinking water was available. Grazing was average quality (*Digitaria eriantha, Cynodon dactylon* and *Panicum maximum*) and sun-cured hay made from these species was fed with molasses and chopped green forage (maize, sorghum and Napier grass). A commercial concentrate was fed consisting of copra, maize grain, rapeseed meal, cottonseed meal, wheat bran, sunflower meal, molasses and an imported mineral supplement. The cows were fed concentrate three times a day, once at each milking and once between, to help increase food intake.

The main barn was a well-ventilated, open-sided building with a high ceiling and was properly orientated in relation to the sun and prevailing winds.

Ticks were controlled by dipping in tanks once a week to prevent anaplasmosis, babesiosis, heartwater and East Coast Fever. Rift Valley fever, actinomycosis (lumpy jaw) and actinobacillosis (wooden tongue) were present in the area, but caused no serious problems. Cattle were tested annually for brucellosis and tuberculosis and vaccinated against foot-and-mouth disease and anthrax. Pink-eye (infectious keratocon-

25

junctivitis) was common. Mastitis was controlled by strict sanitary measures.

Milk yields of first lactation heifers were approximately 3000 kg and the main factors contributing to these levels of productivity were as follows:

1   Feeding low roughage/high quality diets to reduce heat production and stimulate consumption in three separate meals with molasses.
2   Provision of adequate shade in paddocks.
3   Reduction of environmental stress by providing open and airy housing, shade, cool water and by reducing activity of cows.
4   Cooling animals by allowing them to swim in the local river.
5   Strict disease and parasite control.

These measures require good management and high capital inputs. Since milk production is difficult even under the best conditions, high environmental temperatures often cause milk enterprises to be non-viable and are a further constraint to be overcome. Intensive production would not be recommended unless a good price for milk is guaranteed and a high level of management could be provided. A major problem is fertility which is related to feeding, management and the environment.

United Arab Emirates (UAE), Middle East

Intensive milk production from exotic cattle maintained under conditions of high environmental temperatures in the United Arab Emirates commenced in 1970 when 30 purebred Friesian in-calf heifers were imported with two bulls. The area has high temperatures and humidities, low wind velocities and little diurnal variation – all the essentials for a high heat stress factor. The same diet was fed throughout the year. Dry matter and energy intakes were lower than for similar animals in Europe, but the animals maintained good body condition.

Table 2.2   Daily maintenance ration for one Friesian cow in the UAE

Time of day	Feed
04.00	Fresh cut alfalfa (6.8 kg)
07.00	Bran (1.8 kg) + vitamin and mineral supplement (14 g) + dicalcium phosphate (57 g)
11.30	Fresh cut alfalfa (6.8 kg)
16.30	Commercially prepared dairy concentrate (3.2 kg)
17.30	Fresh cut alfalfa (6.8 kg)

In addition, dairy concentrate was fed, 1.8 kg/4.5 kg milk, given in two feeds during milking. Drinking water was available at all times.

Average 305-day milk yields were 3253 kg and 4569 kg for the first and second lactations respectively. These levels were achieved even when animals showed signs of heat stress which occurred regularly. The only alleviation of the climate was the provision of shade. In 1972 a rotating sprinkler was erected in the roof of the cowshed and animals had free access to the spray from 11.30 to 17.00 hours each day. Daytime rectal temperatures rose steadily from approximately 38°C to 39°C over the spring period and then remained between 39°C and 40°C during most of the summer, although temperatures of 41°C were not uncommon at 16.00 hours. This temperature apparently could be tolerated for considerable periods each day in some animals, provided it was cooler at night for three or four hours so that animals could cool down. At the onset of the cooler weather in October yields showed a marked upwards trend which nearly offset the previous depression caused by heat stress.

Many cows did not conceive in the hottest months: cool season breeding was therefore recommended. This was in part associated with the bulls, who had poor libido in the hotter months.

Calves over two weeks of age could tolerate the most extreme conditions encountered, but younger calves showed little thermoregulatory ability and could die if unattended. Usually one or two soakings of the animals' body surface in mid to late afternoon were sufficient to reduce heat stress enough to allow them to survive.

This example in the UAE indicates that milk can be produced using exotic cattle under extreme environmental conditions, but can only be achieved with high capital inputs and good management. Such inputs and demands on management would make milk production of this type beyond the scope of most farmers and the wisdom of attempting this sort of milk production must be questioned by planners and farmers. Even so, large, intensively managed Friesian and Jersey herds have been remarkably successful in Saudi Arabia and the Gulf States.

Summary of characteristics of intensive production systems

- Intensive milk production is a sophisticated form of animal enterprise, requiring high levels of investment and management ability, a reliable source of concentrate food and an assured market for milk at a good price.
- The environment is controlled by good management (i.e. greater inputs of capital, foods, health care, housing and equipment for milk storage).
- The price of milk is critical for the success of intensive systems and must be enough to cover the capital costs involved.

- Many intensive schemes have failed in the tropics. The examples described (see pp. 22–27) indicate the difficulties involved and the likely causes of failure.
- Intensively managed cows should produce a calf once a year to ensure a maximum level of milk output from each cow.
- Intensive production could be complementary to arable farming, but is more likely to compete with arable farming for land. The use of concentrate food inputs for cows could, but need not necessarily, compete for ingredients that humans could consume directly.

Milk collection schemes

The provision of a guaranteed market, transport and storage facilities for milk encourage farmers to increase milk production. Milk collection schemes have been introduced in a number of regions, some more success-fully than others. Collection in itself does not increase production, but facilitates milk off-take. There is a danger that the calf will suffer if farmers take too much milk. If there is no market for milk, farmers are likely to allow the calf to consume all the milk that is surplus to family requirements thus increasing calf growth and revenue from calf sales. The objective of milk collection schemes should be to stimulate increased production by creating wealth with reinvestment in food inputs and health care.

Milk is a bulky product containing more than 80% water and is difficult to transport. It has short storage life and must be consumed immediately unless it is processed to other products (see Chapter 8). Very often milk animals are kept in rural areas away from markets. These factors affect the viability of milk production systems.

A necessary condition for increased milk production is the provision of assured marketing outlets that are sufficiently remunerative to producers. Planners should consider the relative efficiency of alternative milk market-ing systems in terms of costs and marketing margins, product hygiene and quality, range and stability of services offered and stability of pro-ducer and consumer prices. The important objectives of dairy marketing policies are:

- the provision of higher and more stable prices to producers
- assurance of reliable supplies at appropriate prices to urban and rural consumers to assure adequate nutritional standards
- improvement of hygiene and quality.

Governments often tend to favour the establishment of large-scale marketing systems, often state-owned, when other dairy marketing sys-tems may be more appropriate.

> *A milk collection scheme is a private or co-operative marketing system, which provides producers with a guaranteed market at a fixed price. The scheme must be able to store and distribute milk to consumers and guarantee timely payment to producers.*

Examples of milk collection schemes

Two studies, from India and Uganda, indicate the economic, social and political influences on the development of dairy industries as well as the environmental and technical factors involved with milk collection.

There are difficulties associated with achieving balanced and sustained development, even when there is a ready supply of and demand for milk.

India

The example and successful introduction of milk collection schemes in India have stimulated other nations wishing to expand their industries. From 1950–90 output has doubled in India. Out of 180 million cattle and 62 million buffaloes in 1977, there were 55 million milch cows and 31 million **milking buffaloes**. Cow productivity is low (200 kg/lactation), but buffaloes produce greater quantities of milk.

Some 75% of the Indian human population live in villages where average farm size is 2 ha. An extensive infrastructure exists of 600 Key Village Blocks, 130 Intensive Cattle Development Projects, 140 Cattle Breeding Farms, 44 Exotic Cattle Farms and 56 Frozen Semen Banks. The number of crossbred cows is reported to be 5 million and it is intended to develop a national dairy herd of 10 million high-producing cattle and buffaloes.

The history of the dairy co-operative system in India began in 1946 with the establishment of the Anand Milk Union Ltd (AMUL) (or Kaira District Cooperative Milk Producers Union). The Union started with two village milk producers' societies that began pasteurising buffalo milk for the Bombay Milk Scheme in 1948. The organisation grew rapidly and the Kaira Union was the first in India to manufacture milk powder, condensed milk and baby food. The co-operatives ensured a fair and stable price for the producer and the formation of co-operatives into a Union allowed the provision of large dairy processing units and technical services.

In 1970, Operation Flood commenced with the objective of establishing a co-operative structure on the Anand pattern. Milkshed districts were developed and connected by rail to Delhi, Bombay, Calcutta and Madras. In 1970, some 650 000 litres of milk were processed and by 1980 had increased to 2.9 million litres. By 1981, some 12 000 village co-operative milk producers societies had been established in 27 selected milkshed districts. This was expanded by 1984 to 28 174 village producers in 155 milkshed districts linked to markets in 147 towns. The National Milk Grid is being established to link rural milksheds to major demand centres.

It was realised at the beginning of the project that protection from the depressed world market was a pre-requisite for the successful local development of milk production. While a pricing policy aims to make smallholder production remunerative, it does not encourage specialised large-scale dairying where milk animals would compete with humans for food. Constraints to development include lack of food availability, low genetic potential, poor health care, poor performance, inadequate pricing and infrastructure development and inadequate farmer training.

Uganda, East Africa

In Bunyoro District in western Uganda a milk collection scheme was established in the early 1960s. Bunyoro has three pastoral zones: Lake Albert Flats which is a dry bush best suited to beef; the interior *Combretum* plateau which can support one animal per 2 ha; and the Central Hills which are well watered, fertile and suited to intensive milk with two animals per 1 ha. Tsetse and rinderpest were controlled prior to 1960 in the area around Masindi. Previously cattle were excluded and the only source of milk was from goats.

Milk was imported from Kenya until 1960 and sold in Masindi and local areas. Some individuals cycled daily into the town to sell milk from a churn. This milk was derived from local *Nyoro* cattle, which yield about 2 litres/day per cow. In 1962 some Guernsey cattle were imported. At the same time farmers and businessmen were beginning to exploit the potential of milk production. Several attempts were made between 1961 and 1965 to open a milk shop at Masindi, but the venture failed because producers preferred to sell the milk themselves at a higher price. No cooling equipment was available at this stage.

In 1965 a group formed a co-operative milk shop in Masindi where milk from the members could be retailed to raise capital for a separate beef venture. The project received official approval, a shop was provided by the Co-operative Department and a cooler acquired. The existence of centralised cooling and bulking facilities permitted regular contract sales to hospitals, schools, the prison and hotels. The shop was processing 32 000 litres/month in 1968. Supply began to outstrip demand, but there were no facilities for delivery beyond the town. The co-operative used local by-laws to achieve a monopoly for selling milk: as a result, villages near to Masindi had no supply of milk, because local producers could not sell in the villages and local shops could not buy milk from Masindi, even though milk supply was increasing. Once a central sales point with cooling facilities had been established, it became practicable to organise motor transport of milk in bulk.

The scheme had a stimulatory effect on milk production in the area. Prior to the establishment of the shop there had been only two herds of exotic cattle in Bunyoro, but in 1968 there were 1542 exotic cattle in the district. There were severe management and health problems with exotic cattle and many died, but the basis of a dairy industry had been established.

From 1967 to 1970 the scheme consolidated its success. In May 1967 the Dairy Industry Corporation was formed to develop a national dairy industry aiming to eliminate imports from Kenya; and in 1968 the Corporation became a sole buyer of milk. In November 1967 another dairy opened at Hoima with similar success to that at Masindi. In 1970 packed and chilled milk was sold in the towns and milk was marketed from the churn in the countryside. There were 180 licence holders at this time.

Summary of milk collection schemes

- Models of successful milk collection schemes exist which rely on the existence of a profitable market for milk, a means of collecting milk from producers and distributing it to consumers.
- Such schemes are usually based on a central collection point, often with a bulk storage tank and cooling facilities that require a supply of electricity and the ability to maintain equipment.
- Success also depends on the ability of the scheme to guarantee timely payment to producers and a competitive price for milk, the ability to safeguard against adulteration of milk and the maintenance of supply to consumers.
- The growth of milk collection relies to some extent on the entrepreneurial skills of local traders.

Extensive pastoral systems are largely dependent on environmental factors, whereas in semi-intensive and intensive systems environmental influences are modified by the farmer's activities. Increased milk production can be achieved by *either* increasing the number of producers *or* by increasing the output of animals already owned. The former depends on people being in a position to purchase milk producing animals and then to maintain them. The latter depends on the farmers' ability to identify problems with their production system and then to correct them – this might require advice from Extension (Advisory) Services and capital to purchase inputs.

> *To produce and take more milk from the system, any farmer has to put more in, by way of capital, time, effort and labour.*

31

3 Calf rearing

Aims of calf rearing

The productivity of individual cows and whole herds depends on the production of calves. The annual calf crop provides herd replacements, and, through selection of the best calves, can contribute to genetic improvement. In some ways, milk can be considered a by-product of calf production. After calving, the cow must be provided with good husbandry and management to achieve optimum milk yield and to ensure reconception to sustain production year to year. Subsequently the female calf must be reared successfully to maturity.

> *The reproductive cycle provides the momentum of the dairy enterprise.*

Attention to calf welfare begins with the timing of conception and the care of the cow through pregnancy. The ability to control these depends on the calf rearing system. Body condition and nutritive status of the cow in late pregnancy both influence calf birth-weight and survival (pp. 47–48); chapter 3 focuses on calf care from birth.

Methods of calf-rearing

Whatever the management system, to grow well the calf from birth needs the correct quality and quantity of food and protection from disease and environmental conditions (especially high temperatures and heavy rain). The three methods of rearing, (single suckling, restricted suckling and artificial rearing) aim to achieve these objectives in different ways – each having its advantages and disadvantages.

Single suckling

The best method is to rear the calf with the cow for six months or more and for it to suckle freely. This occurs in beef systems and produces high calf growth. The calf receives all the cow's milk and produces large sized calves for meat production. This method is not suitable for the calves of cows whose milk is destined for human consumption.

Restricted suckling

In some situations, milk for human consumption may be more valuable than meat: therefore, some or all milk is removed for human use. The degree of specialisation depends on demand for milk, climate, level of management and availability of alternative calf foods.

A common way in which dairy calves are reared is to allow them to suckle only at certain times of the day. This is known as restricted suckling, or the calf at foot, semi-range, dairy ranching, dual-purpose and partial suckling method. The cow and calf are separated overnight and the cow milked in the morning. The calf sucks for about 1–2 minutes to stimulate let-down and is then suckled after milking. The calf remains with the cow during the day, or in extensive systems is separated when the cow goes to graze. The method can leave the calf hungry if other food is not provided.

It has been suggested that restricted suckling is superior to artificial calf rearing. The former stimulates milk let-down, improves calf growth compared with artificial methods and may reduce mastitis. Cows can show all round best performance (calf growth and milk yield) using this method compared with artificial rearing (Table 3.1). The provision of supplementary food to calves given restricted access to their mothers helps their growth rate. Calves which are fed on a supplement and suckled for 30 minutes after morning and evening milkings can grow

Table 3.1 An example of total milk yields of non-suckled and suckled cows (including that consumed by the calf)

Lactation and milk	Milk yield	
	without calf (bucket-fed)	with calf (suckled)
Number of cows observationed	309	230
Length of lactation (days)	197	262
Milk consumed by calf (kg)	360	524
Saleable milk (kg)	621	1120

faster (up to 0.95 kg/day) than calves supplemented and suckled for 60 minutes in the morning only (0.70 kg/day) or artificially reared by fostering after five days old (0.40 kg/day).

Artificial rearing Artificial teat.

In intensive systems the farmer may decide to sell male calves for fattening for beef, but may wish to rear female calves on the farm as herd replacements. It is usual in more intensive systems to rear the calves artificially. Such artificially reared calves are fed real milk or artificial milk replacer for at least the first 5–8 weeks (and more likely for up to 12 weeks) from a bucket (bucket-fed) or from an artificial-teat dispenser. The method requires good management, without which calves often become ill as a result of poor hygiene.

Calves fed milk or milk substitute with the aid of a bucket should be fed three times a day for the first week and preferably for the first month. Too much milk at a feeding can cause diarrhoea, as will milk that is too hot or too cold. Optimum temperature is 35–38°C but systems that use cold milk are possible if the temperature of the milk is constant and does not fluctuate from day to day. In young calves milk by-passes the reticulo-rumen via the oesophageal groove, a fold of skin in the rumen wall forming a channel from the oesophagus to the abomasum, as a reflex action to the stimulus of suckling or stimuli such as bucket feeding.

Examples of calf rearing systems

Calf rearing in pastoral systems

Similar calf rearing systems have evolved in pastoral systems throughout the dry zones of the world providing time for the cow to graze each day, for the calf to suckle and for milk to accumulate for milking.

Such methods are illustrated by the Fulani pastoralists. At night, herds are confined near to the homestead (*ruga*) with the adults tied in pairs according to age or size. In some areas the adults are not tied, but are confined in a corral (fenced enclosure). The calves are tied separately in order of age by neck loops along a rope which is tied between two posts. Milking is usually carried out in the morning when calves suck briefly to induce milk let-down. They are then tethered to the dam's foreleg until milking is completed and then suckle *ad lib.* until the herds leave for grazing (Fig 3.1 and 3.2). Very young calves are kept near to the homestead during the day for 3–4 weeks after birth before allowing them to follow the cows during the day. These are provided with water by women and children once a day and suckle their dams again after the day's grazing before being confined at night away from the cows. The

34

Fig 3.1   *Fulani cow being milked with a calf at foot*

Fig 3.2   *Fulani calves remaining in the* ruga *while the cows are away grazing*

older calves accompany the cows and are separated only during the evening or may be herded separately away from the breeding bulls to avoid early conception. Milking is carried out by hand by men and boys who regulate the amount left for the calf. The milk then becomes the property of the women. Milking continues until the cow is nearly dry, but may stop earlier if a cow is pregnant.

Dairy ranching

Restricted suckling is also the practice in dairy ranching systems (see p. 15) in which dual-purpose cows provide milk for human consumption. Calves are separated overnight and suckle for one minute prior to milking in the morning and then remain with the cows for the rest of the day.

Smallholder production

Calves born to cows kept in settled village systems are reared on the cow by variations of the restricted suckling method. Cows may be herded around the village, tethered nearby or stall-fed and the calf would usually stay with the cow during the day and be removed at night if the cow was being milked. Alternatively, calves may be coralled during the day while cows are grazing or kept near to the house for protection.

Intensive production

In intensive systems the farmer must decide

(1) to rear calves on the farm, to sell or to give them to someone else to rear;
(2) to rear them either naturally on the cow or by artificial methods using milk substitutes and other calf foods.

Suckled calves which also graze, can achieve relatively high rates of growth, but if the transition from a milk diet to a roughage diet at weaning is too rapid, this can cause severe checks to growth. Tropical climates allow calves to be put out to pasture aged 3 months, when their rumen is inadequately developed for roughage diets. Ideally, it would be better to raise artificially reared calves away from pasture for as long as possible (6–9 months), but economic considerations usually force farmers to choose a more extensive method of calf rearing, with some inclusion of grazing pasture (see p. 42).

> *To have the best chance of survival, a calf should be born into the best possible environment and at the most favourable time of year.*

Care of calf at birth

The season of calving is often beyond the control of the farmer; it depends on the time of conception which is determined by nutritional and other stimuli. If a calf is born in the dry season there is less likelihood of it becoming chilled from rain than if it is born in the wet season. Chilling can reduce resistance to infections and losses often occur at this time. It is more likely that calves will be born at the beginning of the wet season, after conception at the end of the previous rains. Cows conceive at this time because they are in good body condition as a result of a high level of grass intake during the rainy season.

Preparation for parturition

At 12–24 hours before calving the cow becomes uneasy and moves away from the rest of the herd. Signs of calving include enlargement of the vulva, distension of the teats and udder and loosening of the ligaments at the side of the tail-head. If possible the cow should be isolated when calving is imminent and kept in a quiet and protected place.

In pastoral herds, this is difficult and calves are often born when cows are out grazing during the day; once the herd returns to the overnight base, the cow may be left for the next 24 hours with the calf and fed on cut grass. In settled systems cows can be provided with greater seclusion and in intensive systems special provision is usually made for cows at calving; a special stall or calving box may be provided with a concrete floor and cemented walls which are easy to disinfect before and after calving. In village systems, cows are often kept in poorly constructed night kraals which are unhygienic and muddy – such an environment is not suitable for cows to calve into; better provision should be made for cows at calving.

The cow should be left to calve alone and only if she shows signs of having difficulty should the farmer intervene. The normal presentation of the calf is with the forelegs pointing forwards with the nose between them (Fig 3.3 and 3.4). Difficult calvings, particularly in young heifers and small cows may occur when a large calf is born, which causes difficulties at calving (dystokia).

Care at parturition

On calving, the fetal membranes should be removed from the calving area or pen floor, mucous should be removed from the nose and mouth of the calf and, if possible, the navel dressed with tincture of iodine (or copper sulphate or carbolic solution) to prevent local infection and to

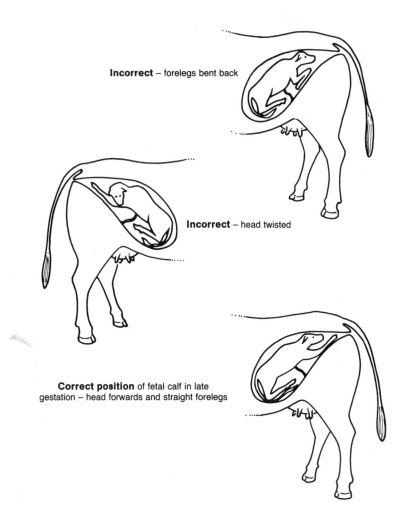

Incorrect – forelegs bent back

Incorrect – head twisted

Correct position of fetal calf in late
gestation – head forwards and straight forelegs

Fig 3.3 *Positions of the fetal calf*

help the umbilicus heal quickly. The fetal membranes should be voided
from the cow in the first 12 hours after calving. The cow will lick the
calf dry and the calf should suckle within the first 2–3 hours. During the
first 4 days the calf will suckle eight or more times a day.

Colostrum

The milk produced by the cow for the first 3–4 days is called colostrum.
This contains antibodies immunoglobulins G and M (IgG, IgM) from

the cow's immune system: these confer passive resistance to many infections. No transplacental transfer of antibodies occurs. Colostrum has a high nutritive value and its laxative properties stimulate the removal of the calf's first faeces (meconium). The yellow colour of colostrum is due to the high carotene (Vitamin A precursor) content, derived from green vegetation. Colostrum also provides Vitamins B, D and E. The optimum time for absorption of antibodies through the calf's small intestine is the first 6–8 hours and absorption is aided by the presence of the dam. Hence the calf should be left with the cow for at least 12 hours.

If the calf does not receive colostrum during the first day, its chance of survival is greatly reduced as the intestines take-up the antibodies *only* during the first 12–24 hours. It should be standard practice to drench the calf with up to 2.5 kg colostrum within the first 6 hours after birth and again within the next 6 hours, even if the calf is seen to be suckling. This will ensure that the calf receives enough colostrum and will protect against early infection.

Ideally the calf should be kept with the cow continuously after birth, but if the cow is to be milked, this is not possible. In pastoral or settled extensive grazing systems the calf may be left alone during the day while the cow goes to graze.

Fig 3.4   *Correct presentation of the calf for parturition*

Front feet and muzzle of calf appearing first

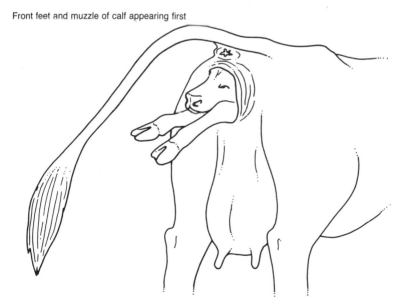

Bucket rearing

In intensive systems the farmer may prefer to wean (to stop the calf suckling the cow) at birth so that the cow can enter the milking herd. The calf may be reared by feeding milk or milk substitute from a bucket. The calf should suckle for the first 12 hours and then be removed from the cow. If the calf remains too long with the cow, it will be less easy to teach it to drink from a bucket.

This system has many disadvantages and bucket-reared calves usually do not grow as well as suckled calves. They are less well fed and more prone to disease, mainly because of poor hygiene and stress. For example, suckled White Fulani (Bunaji) calves reared at Livestock Breeding Centres in Kano State, Nigeria, weighed 69.9 kg at 12 weeks, compared with 54.4 kg for bucket-fed calves.

Calf feeding

At birth the calf requires up to 3 kg of milk a day depending on size, but may be able to drink more. In newly born calves the rumen is not developed and its process of digestion is similar to that of a monogastric animal, such as pigs and humans. Milk does not stay in the rumen, but by-passes it via the oesophageal groove. The calf makes optimum use of milk before its rumen starts to function and calves can be expected to make the best growth at this time.

> *Milk is the best food for calves and every opportunity should be taken to provide the calf with as natural a diet as possible.*

Some mineral deficiencies may occur in calves kept on milk for too long and calves require some solid food from an early age to provide vitamins and minerals (e.g. magnesium and iron) deficient in milk. The vitamin status of the calf depends on its own body stores, the amount received in milk and feed, synthesis of vitamins in the rumen and tissues, the rate of utilisation by the calf (which can be increased by disease) and the rate of absorption (which can be reduced by disease).

Whole milk

In single and restricted suckling, the calf receives whole milk from the cow. In systems using buckets, the calf may be bucket-fed with milk from the cow, but feeding whole milk (which has been milked out of the cow

Table 3.2 Recommended amounts of whole milk and concentrate for dairy calves up to 12 weeks of age   *✗ see cover* (handwritten)

Age (weeks)	Whole milk (kg/day)	Concentrate meal (kg/day)
+ 1 *grass + water* (handwritten)	3.0	—
2	3.5	—
3	4.0	—*
4	4.5	—
5	5.0	0.10
6	5.0	0.20
7 ~~milk forms~~ (handwritten)	5.0	0.30
8	4.0	0.40
9	3.0	0.70
10	2.0	1.00
11	1.5	1.25
12	0.75	1.50

* Hay and meal are offered from week 3. *of cows b worm in festation give 1-2 BCD a 300ml cod liver oil* (handwritten)
8 wks to pasture – ahead (handwritten)
Example from Zimbabwe for a 25–30 kg calf at birth, receiving 350–400 kg milk over 12 weeks, making gains of 0.5–0.75 kg/day and weighing 70–100 kg at 12 weeks.

Source: Oliver (1987)

by hand or machine) is expensive and other methods are preferred. The feeding of milk should begin at a level equivalent to 8–10% calf body weight per day, increasing gradually to 5 kg or more daily (Table 3.2).

Milk replacers are made from lower grade substitute constituents, such as vegetable oils and proteins, and resemble cow's milk in composition, but are never fully equivalent in nutrients.

Bucket feeding can cause digestive disorders if hygiene is poor. Buckets and all equipment must be cleaned well between use to avoid this. If calves suffer from diarrhoea (**scours** in calves), it may be necessary to dilute milk and it is always advisable to dilute milk containing a high percentage of fat before it is fed to calves.

If the calf is given the opportunity to eat solid food, rumination will start at age about 4 weeks. Solid food stimulates rumen development and this is important for dairy calves who in later life must be able to eat enough to maintain milk production. Calf starter rations, gruels and mashes should have high protein (14–16% crude protein) and energy contents (8–10 MJ/kg DM) (see chapter 6). Small quantities of green material may be offered after the first few weeks and good quality hay can also be given. Urea (1%) can be added in solution to the gruel or wet ration after about age 1 month, plus mineral and Vitamin A supplements. Calves should have access to clean water at all times.

Calf rations can vary according to the feeds available. An easily pre-pared food is 1 kg cooked maize mixed with 10 kg water and fed at a rate of 0.9 kg/day at week 10 increasing to 4.5 kg/day in week 40.

Early weaning

Milk substitute is usually expensive relative to other calf foods because it is manufactured and often imported and the price for calves may not justify high rearing costs. These costs can be reduced by replacing milk with dry meal, hay and water. This also reduces the risk of scours caused by bucket feeding. The calf must be fed milk for the first 3 weeks at a level equal to 8–10% of body weight, but thereafter could be weaned-off milk if required. This is not recommended (since the calf's growth rate will undoubtedly be reduced) except under exceptional circumstances where milk is required for sale and receives a high price. Early weaning on to dry diets requires high levels of husbandry and can have quite disastrous effects if husbandry is poor, because calves will be under-nourished and prone to infection. During the first 3 weeks, *ad lib.* concen-trate and chopped hay (10% of ration) can be given. Concentrate 1.0–1.5 kg will be eaten by day 24. Days 25–150, no milk need be given and calves will eat up to 4 kg of concentrate feed fed *ad lib.* depending on body weight. After day 150, the amount of concentrate fed should be reduced to a level that maintains steady growth. After weaning, the calves can be kept together in a unit in which they can feed by themselves. They should be drenched against tapeworms at day 30, against roundworms at day 60 and disbudded with a hot iron at day 30. After day 150 the calves could be put out to pasture and if so, they should be drenched against roundworms at the beginning, middle and end of the rains. Calves reared in this way should gain 0.75 kg/day to 150 days of age.

Calves fed on dry diets will require a constant supply of water and will be able to drink 10% of their body weight of water each day and for young calves, this figure may be 20%.

Calves at grass

Extensively managed calves in pastoral systems and settled village herd-ing systems are introduced to grass feeding at an early age. In intensive systems it is possible, though there are disadvantages, for calves to be raised outside on pasture from 2–3 months of age. In warm climates the requirement for protective housing for calves is less than in temperate regions, except for the need to provide shade. This makes management easier, since it puts a smaller demand on housing facilities.

> *Growth rates of calves fed on grass are lower than growth rates that would be achieved if the calf stayed on a milk diet, because the rumen takes time to become adjusted to roughage diets and these are often poorer in quality than milk diets.*

Calves have a potential for rapid growth and consequently have a high protein demand which is incompatible with the roughage available in the tropics for much of the year, even in intensive systems. Problems of achieving a balanced intake of nutrients are less likely to occur in suckled calves at grass. Tropical pasture species grow rapidly and are most palatable in the young vegetative growth stage. They quickly mature becoming hard and woody and more unpalatable to the calf. When calves are no longer fed on milk and have to rely on pasture, their growth declines and they usually require supplementary feed to avoid weight loss. Supplementary feed is usually not provided in pastoral and smallholder systems. In intensive systems it would be preferable to rear calves inside until they are 6–9 months old if management would allow, because this would allow optimum milk/concentrate food to be provided. Extensively managed calves in pastoral systems and settled village herding systems are introduced to grass feeding at an early age (see p. 42).

Routine health care

A healthy calf is active, has a smooth, shiny coat, bright eyes, good coordination and shows no jerky movements. Calves which stand with their head lowered, have rough coats and pale mucous membranes, give signs of faecal contamination around the tail and pot bellies are showing signs of ill health.

Calf mortality is usually highest just after birth, but with good care and husbandry, losses can be reduced to low levels. There are two major problems:

(1) scours in the first 3 weeks and
(2) respiratory diseases in the second 3 weeks.

The importance of colostrum for good calf health (see p. 38) cannot be over emphasised.

Calves are sensitive to levels of hygiene, nutrition and the quality of housing. They are prone to chilling (causes pneumonia) or may suffer heat-stress in climatic extremes. In addition to navel ill (see p. 44) and calf scours (see p. 44), calves are susceptible to respiratory diseases, such as pneumonia caused by bacteria (*Pasteurella* spp.) and viral agents. Attention must be given to ventilation. The farmer should also be aware that calves are inquisitive and will lick any object in their environment. This can lead to the consumption of poisonous substances. Calves must be wormed regularly through the wet season and at the beginning of the dry season to avoid the detrimental affects of worms on health and growth. Disbudding and castration are routine operations which, if they are to be carried out should be done early in the calf's life.

Five principles of control and prevention of infectious disease in calves are as follows:

- Remove causal agents from the calf's environment.
- Remove the calf from infected environments.
- Boost non-specific calf immunity by ensuring adequate colostrum intake.

- Boost specific calf immunity by vaccination of the cow to increase required maternal antibodies in colostrum (calves do not respond to vaccines until 2 months after birth).
- Prevent stress, a factor known to cause pneumonia to develop from the ever-present *Pastuerella* bacteria.

All female calves should be inoculated with S19 vaccine against brucellosis (contagious abortion) at 4–8 months of age.

Common health problems of calves

Navel ill

Navel ill (or joint ill) is caused by infection via the navel soon after birth and can be avoided by dressing the umbilicus with antiseptic and maintaining good hygiene at parturition (see p. 37). Navel ill is a septicaemia: infection of the umbilical region occurs and joints swell. Bacteria commonly associated with the disease are *Corynebacterium pyogenes*, *Escherichia coli*, *Fusiformis necrophorus* and *Staphylococcus* spp.

Calf scours

A general term for calf diarrhoea, scours can be a problem in the first 3–6 weeks after birth if calves are poorly managed. Outbreaks can occur in the first 6 weeks with progressive dehydration and weakness followed by death. The causal agent is usually *Escherichia coli* (white scours), a species of bacterium normally present in the intestines of cattle. Other organisms which cause diarrhoea include *Salmonella* spp., Rotavirus, Coronavirus, coccidia and increasingly in the tropics, *Cryptosporidium* spp. Dietary abnormalities and coarse feeds may act as predisposing factors. The bacterium *Clostridium perfringens* causes enterotoxaemia in calves 7–10 days old.

Cows can be vaccinated at 6 weeks and again 2 weeks prior to calving and then annually. All infected calves should be isolated and treated with fluids. A mixture of glucose (1 cup), sodium chloride (salt) (1 teaspoon), bicarbonate of soda (1 teaspoon) and potassium permanganate (¼ teaspoon) dissolved in water (2 litres) is a suitable treatment. Calves should receive 4 litres/45 kg body weight over 24 hours divided into four doses. Dams can be vaccinated for *E. coli*, Rotavirus and Coronavirus 2–5 weeks prior to calving to increase maternal antibodies.

Helminths

The common worm infestations of calves can be divided into three groups:

(1) stomach/intestinal worms (parasitic gastro-enteritis);
(2) tapeworms; and
(3) liver fluke.

Gastro-enteritis is caused by Trichostrongylidae worms in the abomasum (e.g. *Haemonchus contortus*) and small intestine. Symptoms of parasitic infection include poor body condition, starring coat and pot belly. These signs are common in calves born at the beginning of the wet season. A number of anthelmintics are available, some of which will also control other forms of worms. Calves must be wormed regularly, particularly at the beginning of the rains.

Calf pneumonia

Pneumonia is a common and troublesome condition of confined calves reared in close proximity with each other. It is not common in calves reared in extensive systems. Pneumonia is caused by bacteria *Corynebacterium pyogenes, Escherichia coli, Actinobacillus actinoids, Fusiformis necrophorus, Pasteurella haemolytica* and *P. multocida*) and viruses. The condition can be prevented by providing clean, airy and uncrowded conditions.

Calf housing

In many production systems, no housing or shelter is provided for calves of adult cattle. At the least a simple shelter is recommended for newly born calves. Rapid growth is desirable to help young calves over the period when they are most susceptible to disease and digestive upsets. During the first 10 days of life the calf's temperature control mechanism is poorly developed; calves easily suffer from heat stress, in the heat of the day, and chills, if they get wet in the rain.

> *Stress from exposure to heat and wet, predisposes the calf to further infection and reduces growth rates. A good shelter made of poles and thatch will help to overcome these problems and reduce calf mortality rates.*

Calves reared artificially should be housed separately from birth to about 3 months of age. This prevents calves sucking each other, swallowing hair and ensures they all receive their fair share of food.

Buildings can be of simple designs, but must be kept thoroughly clean and disinfected. Closed, fixed shelters in hot, wet climates can easily become sources of heavy infection. Pens must be dry and protect the calves against extremes of temperature while providing adequate light and space. Ventilation is necessary to remove excess carbon dioxide and ammonia. Provision of ventilation should present few difficulties in the tropics. Protection against sidewinds and rain is necessary in high rainfall areas. Floors of hard, impervious material are easily washable. Raised, slatted floors provide ventilation and improve the state of cleanliness.

The calf-house can provide protection against some diseases, but badly designed houses may increase the risk of disease. The risk of pneumonia is increased by poorly ventilated houses because of the increased risk of cross infection and the concentration of bacterial and viral agents. The risk of calves being infected with helminths is increased if they are kept in dirty paddocks or yards.

Fig 3.5   *Calves on slatted floors, in Malaysia*

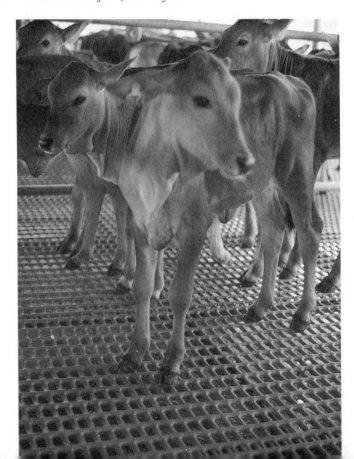

Suitable calf-houses include wooden or metal crates, brick pens, hutches or kennels made from straw bales. The latter have the advantage that they can be burnt after each calf has been reared. Slatted floors help to keep houses clean and can be hosed down daily, though deep litter may be more comfortable and warmer for young calves. Litter can be allowed to build up, but it must keep the calf clean and dry and should be changed every 7–14 days.

Calves can be reared together from 2–3 months of age and at that stage should be allowed 5.5 m² each. Calves of similar sizes should be kept together in batches.

Calf growth

At birth, calves vary in weight according to breed and sex (Tables 3.3 and 3.4), though their size and body conformation also depend on the size and condition of the dam. The birth-weight of calves should be approximately 7% of the dam's weight. Ideally the cow should have a body condition score 3.0–3.5 (see p. 68) at conception and this condition shuld be maintained throughout pregnancy.

Table 3.3   Average calf birth-weights of different breeds

Breed	(kg) Birth Weight		Country
	Female	Male	
White Fulani	21.0	22.4	Nigeria
Mpwapwa	27.8	–	Tanzania
Brahman × White Fulani	24.0	–	Nigeria
Friesian × White Fulani	27.5	–	Nigeria

Calves from birth will gain up to 1 kg/day depending on breed and birth-weight, but lower rates would normally be achieved. Calf growth is related to birth-weight and it can be assumed that calves will grow at a rate of 1.5% of their body weight per day. Heavier calves will put on more weight daily than lighter calves (Table 3.5).

Different breeds exhibit different growth patterns according to their genetic make-up, but more importantly growth is affected by environmental factors such as nutrition and husbandry. The differences between breeds and management system in terms of calf growth are illustrated by the results of work carried out in Nigeria (Table 3.4) on indoor and outdoor rearing of bucket-fed White Fulani and White Fulani × Friesian calves. Crossbreds gained more than pure White Fulani (0.60 kg/day compared with 0.42 kg/day) and confined calves gained more than calves kept on pasture (0.52 kg/day compared with 0.45 kg/day).

Table 3.4 Weight of different breeds of heifers at different ages

Breed	Weight (kg)						Country
	Age (months)						
	6	12	18	24	36	48	
White Fulani	–	–	–	204	–	–	Nigeria
Bunaji – male	140	204	258	326	–	–	Nigeria
– female	122	188	234	279	–	–	
Mpwapwa	–	–	161	–	–	–	Tanzania
Brahman × White Fulani – female	109	180	216	279	376	401	Nigeria
Friesian × White Fulani	124	179	245	303	378	435	Nigeria

Growth also depends on the time of year (season) when the calf is born. Calves born at the end of the wet season or beginning of the dry season will have dry, mature and fibrous herbage available and thus will not grow as well as calves with access to better herbage in the wet season. Growth rate can be as low as 200 g/day. Calves will keep suckling until they are forcibly weaned or the cow dries off. This can contribute to delayed conception and therefore cause long calving intervals, but cannot be avoided if the calf is to receive the best value from the cow's milk at times of the year when other food is scarce and poor quality. If calves were weaned in extensive systems at earlier ages, they would undoubtedly suffer nutritionally.

Table 3.5 Weight of calves at 91 days of age in relation to birth-weight and daily percentage weight gain

Birth weight (kg)	Weight gain (%/day)			
	0.875%/day		1.5%/day	
	91-day weight (kg)	mean gain (kg/day)	91-day weight (kg)	mean gain kg/day
20	44	0.27	78	0.64
25	55	0.33	98	0.80
30	67	0.40	117	0.96
35	78	0.47	137	1.12
40	89	0.54	157	1.28
45	100	0.60	176	1.44
50	111	0.67	196	1.60

Source: Roy (1980)

4 Heifer rearing

Female calves are called **heifer calves** and after weaning are called **heifers**. After they have calved for the first time, they may still be called **first calving heifers**.

Good heifer care is important to any dairy enterprise. The amount of milk that a cow produces in her life depends on the number of lactations completed, in turn depending on the number of calves born. If the first calving occurs at too early an age this may cause stunted growth, which may reduce lifetime production. The period between weaning and first calving should not be seen as an unproductive period, but as the foundation for the animal's productive life.

General aspects of heifer rearing

Heifer calf management to 150 days is as outlined for calves (see p. 32). Later management depends on the system and the season of birth. In extensive systems, such as pastoralism and small-scale production, calves will grow more slowly than in intensive systems, since conditions will be less favourable. Any form of supplementary food would be beneficial to such calves.

> *In natural grazing systems, calf growth receives many checks after weaning due to seasonal changes in the quality and quantity of herbage production. In intensive systems, the objective is to even-out natural fluctuations of food supply and to provide a constant supply of food.*

The objective of heifer rearing must be to achieve the best growth for the economic circumstances and capital resources of the farm, particularly since the most efficient use of food for growth occurs in younger animals and better feeds are used with a greater efficiency for growth than for milk production.

Calves born in the wet season

Calves born in the wet season will enter the dry season at about 5 months of age. In extensive systems, such calves will be dependent on natural vegetation (grasses and browse), but if possible should receive supplementary food, such as maize bran or sorghum bran, in addition to milk from the dam.

In intensive systems, calves could be placed in small paddocks of improved pasture such as star grass (*Cynodon* spp.). Care should be taken to avoid hydrocyanide poisoning from new grass which causes sudden death; this form of toxicity is common in both East and West Africa. At 5 months of age when calves enter the dry season, they will eat 3.0 kg or more of concentrate if given the opportunity, but will need an additional source of roughage, such as cut grass, tree leaves, hay or silage. As grass becomes available at the beginning of the following wet season, the concentrate should be reduced over the first month to 1 kg and cut grass or conserved fodder discontinued once the amount of grass growth permits. The concentrate should be adjusted according to heifer growth and condition, remembering that growth should be steady. During the second dry season the equivalent of 4.5 kg concentrates should be given daily, together with forage, such as silage (9 kg), or other better quality roughage, such as cut grass (4 kg). These quantities should be given until about 6 weeks before calving (at 24 months of age in an intensive system, but probably nearer 3½ years in extensive systems). The precalving heifer should be well fed and should receive the equivalent of 5.5 kg concentrate and 10 kg fresh cut grass or 23 kg silage daily.

Calves born in the dry season

Calves born in the dry season will attain an age of up to 5 months before the wet season. Supplementary foods such as concentrate and hay should be provided during the dry season. The amount given can be reduced over the period of one month at the beginning of the rains, but the amount will depend on pasture growth and the heifers' condition. The feeding of hay can be discontinued once the animals have settled to the consumption of grass. Subsequent management is as described for calves born in the wet season.

The maintenance energy requirement of a 200 kg heifer is approximately 27 MJ ME/day (see chapter 6) and depending on quality of diet, the additional requirement for liveweight gain is 12–14 MJ/day for a gain of 0.50 kg/day. Such an animal would have to be fed up to 40 MJ/day and this would require an energy concentration in the diet of approximately 8 MJ/kg DM. Heifers allowed access to good quality grass could consume sufficient food to achieve this energy intake. Energy requirements for pregnancy increase up to 20 MJ/day in month 9. If the heifer is growing at a rate of 0.5 kg/day in the ninth month of pregnancy, her requirement would be in the order of 60 MJ/day. This puts greater demands on nutrient supply and an energy concentration of nearer 12.5 MJ/kg DM would be required. This cannot usually be provided by pasture alone (even good pasture) and some form of concentrate supplement would be required.

Puberty and age at first service

Maturity in ruminants is weight dependent as well as age dependent, which means that calves of the same age may be at different stages of maturity depending on their growth. The onset of puberty which is the age at which the **oestrous cycle** begins in heifers, depends on weight and rate of growth. Puberty occurs at approximately 40% mature weight in *Bos taurus* (humpless) breeds and heifers are served at 55% mature weight at 15 months or over. *Bos indicus* (humped) cattle reach puberty at 60% mature weight.

The best age for the heifer to be mated (and hence to calve) depends on her weight.

A steady rate of growth is desirable for the growing heifer as this leads to better life-time performance when she becomes an adult. Steady growth from birth to first calving is required, rather than good growth to weaning followed by a reduction of growth when the calf is weaned and first put out to pasture.

The optimum growth rates after weaning in temperate regions are 0.4–0.7 kg/day depending on breed. To achieve these levels of growth, diets must contain at least 12.5 MJ ME/kg DM and 14% CP (crude protein) on a fresh weight basis. Target weights for first service depend on breed, but should be higher than that at which the heifer achieves puberty and first oestrus. The range and variation in expected target weight of animals reared in intensive systems in the tropics depends on the management and nutritional levels (Table 4.1). Different breeds reach a suitable condition for first service at different weights and different ages.

If these targets cannot be met, this does not mean that the enterprise will be unprofitable. A low age at first calving is desirable because it reduces land and forage requirements for replacement stock and reduces the overall maintenance costs for young stock. A lower age at first calving

Table 4.1 Target weights for first service of some dairy cattle breeds

Breed	Target weight (kg)	Age (months)	Mature weight (kg)	Country
Ayrshire	285	17–22	480	Kenya
Friesian	320	18–23	540	
Jersey	260	14–18	370	
East African Zebu	270	24–26	320	
Jersey	225–250	15	–	Zimbabwe
Friesian	340–360	15	–	
Friesian (Purebred)	400	19.5	–	Nigeria
Friesian (¾)	400	21	–	
Friesian × White Fulani	300	22	–	
White Fulani	350	40	–	

also leads to faster genetic improvement, because this will allow a cow to have more calves and this will allow more selection (chapter 7).

Once the heifer is served and has conceived, she enters a more critical stage when it is important that her feeding is correct. Growth of body tissue must continue and the fetus must develop. This involves the laying down of the fetal membranes and in later pregnancy, the development of mammary tissue. The heifer must grow and also achieve an adequate body condition to support lactation.

Mating the heifer

For details of maintenance of heifer and cow fertility and oestrus detection, see chapter 5. Artificial insemination (AI) is not commonly used in many tropical milk production systems: normal (natural) service using a bull is practised. If normal service is used and the bull runs with the cows, there will be no need for the farmer or stockman to worry about oestrus detection. When AI (p. 126) is used, the precise time of oestrus must be detected for the insemination to be carried out at the right time.

The bull should be checked for campylobacteriosis (*Campylobacter fetus*) formerly known as vibriosis (*Vibrio fetus*), trichomoniasis (*Trichomonas fetus*) and infectious bovine rhinotracheitis (IBR) and other venereally transmitted diseases (passed from bull to heifer during mating) which cause infertility.

The bull can be run with the heifers or when the heifers have been observed to be in oestrus taken to the bull pen or bull holding centre. Since the milk potential of heifers has not been determined (and hence it is not known whether the daughters will be potentially good replacements for culled cows), it is often preferable to mate heifers to a small-size beef breed (if this is feasible). This will result in the calf being smaller and will cause fewer calving difficulties.

Age and weight at first calving

The age and weight of first calving will vary with the system, but often will not be controlled as well as wanted for optimum production. The likelihood is that heifers will be mated earlier than desirable. Heifers in intensive systems can calve at 24 or 30 months; target weights to achieve this are shown in Table 4.2. Heifers which calve at 2 years will have completed a full lactation by the time other heifers calve at 3 years of age. Under extensive systems of management in which animals grow more slowly, a later date of first calving would be preferable to allow the heifer to grow to a better target weight.

Table 4.2 Target calving weights for age for heifers of large-size breeds*

Age (months)	Calving at 24 months		Calving at 30 months	
	Target (kg)	LWG/day (kg)	Target (kg)	LWG/day (kg)
0	42	0.72	42	0.60
5	150	0.66	132	0.50
15	350	0.60	282	0.50
24	512	–	417	0.50
30	–	–	507	–

* Heifers of small-size breeds weigh about two-thirds of the figure given.
LWG = liveweight gain

5 Husbandry and management of milk cows

Cows need to be kept in stress-free surroundings with as little variation as possible in environment and feeding regime. In such situations, cows will maintain good health and body condition, produce a calf regularly and give optimum milk yields. Husbandry and management of milk cows includes:

- general aspects of husbandry practices
- environmental control and housing
- health care
- body condition
- management of lactation
- management of fertility.

Husbandry of milk cows

> Good husbandry provides the best care and conditions so that the cow can produce well.

The farmer can do many things to provide good conditions for milk cows. The first is to provide a safe and hygienic environment. Attention must be given to cleanliness and hygiene around the farm, the removal of pollutants and dangerous objects from the animals' environment, the quality of drains and waste disposal systems, the removal of dung and soiled bedding, the provision of good housing, the routine disinfection of housing to remove parasites and reduce risk of infections, the disinfection of houses before new stock are introduced and the maintenance of clean milking utensils, milking machines (if used) and other equipment.

Secondly a regular husbandry routine should be maintained so that animals are not subjected to unpredictable and unsettling changes.

Stress

Poor conditions and a lack of routine lead to stress and ill health. Inhospitable climate, poor nutrition, bad handling, poor housing, irritation by parasites, minor ailments and unhygienic conditions can cause stress. Stress can cause further ill health by reducing resistance to pathogens and normally avirulent disease organisms present in the environment. Mastitis, for example, can be caused by organisms normally present in the environment, which infect the udder when the animal's resistance is reduced.

Stress can also be caused by health care measures, such as dipping against ticks and tick-borne diseases, and vaccination. These can cause stress directly or indirectly as a result of the grazing time lost. The maintenance of good environmental conditions, the control of adverse effects of climate, the provision of good nutrition and attention to disease prevention measures all help to reduce stress and promote good health.

Routine husbandry practices

Animals should be treated to the same routines each day. They should be fed, watered and milked at the same time and allowed to graze at the same time and for the same duration each day. In intensive systems, clean drinking water should be available at all times, both where the animals graze and in the stalls. If food is changed, this should be done over a period of 4–5 days to allow the animals to adjust their digestive processes. If calves are allowed to suckle, this should be done at the same time each day.

General hygiene and animal safety

The area where animals are kept in stalls or tethered should be kept free of dangerous hazards such as nails, wire, broken glass, plastic bags, holes in the ground and stones. Troughs, feeding apparatus and the areas around these should be cleaned daily. Wet or marshy areas should be fenced or drained and leaking water troughs and pipes repaired. Bedding and fodder which are soiled by urine and dung should be removed and new bedding provided regularly. Poorly stored straw, conserved forages (hay and silage) and concentrate feeds may harbour harmful toxins or moulds. Foods stored in poor conditions may go mouldy and cause digestive upsets. Removal of dung and soiled bedding helps to reduce flies around the farm. Biting flies and ectoparasites cause stress by worrying the animal, sucking blood and causing secondary infection of wounds. Houses can be disinfected using insecticide or acaricide applied with a hand-spray to combat flies, ticks, mites, fleas, lice and termites. Some parasites are more important in the wet season, while others are present all the year round. Severe infestations are caused by poor husbandry, overcrowding and the build up of dung.

Attention to signs of stress

Signs of stress and ill health includes loss of appetite, recumbency, laboured breathing, restlessness, coughing, salivation, increased temperature, nervousness, reduced daily milk yield, discharge from the vulva and changes in the textures of dung. At certain times of the year animals may suffer from heat stress. Signs of heat stress include refusal of the animal to lie down, huddling (surprisingly), body splashing, high respiratory rates, high rectal temperature and open-mouth breathing with the head extended, elbows turned out, tongue protruding and salivation. People caring for stock must spot the signs of ill health or deviations from normal health and attend to the stock before their health deteriorates. These signs can be used to predict problems. **Good** husbandry personnel know what is happening to their stock; **excellent** ones know what **will** happen.

Healthy animals are alert, active, aware of their surroundings, have bright eyes with no signs of discharge, the tail constantly moves to counter flies, they walk easily with sure steps, have smooth shiny coats and breathe and ruminate regularly.

Inspection of animals

Routine inspections facilitate good husbandry and disease prevention. Records of any signs of ill health should be kept for the veterinarian or animal health assistant.

The age of an animal can be found by looking at the **teeth**. The first teeth (**milk teeth**) remain up to two years (Fig 5.1). Adult incisor teeth erupt in pairs and animals are called two tooth, four tooth, six tooth and full mouth depending on the adult teeth present. Once all teeth have erupted the animal's age can only be guessed, but the amount of wear on the teeth gives an indication.

A good indicator of the animal's state of health is its **rectal temperature**, which is normally 38.5–39.5°C. Some diurnal variation can occur if ambient temperatures are high. A diseased animal will usually have a raised temperature. In the terminal stages of a disease the temperature will fall to sub-normal level. Increases in the pulse rate (normally 60–80/min in an adult and 100–120/min in a calf) and respiratory rate (normally 10–30/min in an adult and 20–45/min in a calf) are signs of stress and active infection.

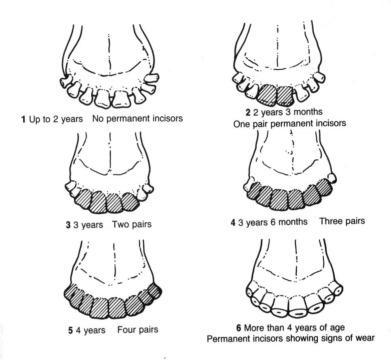

1 Up to 2 years   No permanent incisors

2 2 years 3 months
One pair permanent incisors

3 3 years   Two pairs

4 3 years 6 months   Three pairs

5 4 years   Four pairs

6 More than 4 years of age
Permanent incisors showing signs of wear

Fig 5.1   *Eruption of teeth in Zebu cattle*

The visible **mucous membranes** should be inspected, by looking at the conjunctiva (the inner part of the eyelid), the gums and the lips of the vagina. The mucous membranes usually are rose pink and moist. Pale mucous membranes indicate anaemia which arises through the loss or destruction of red blood cells, commonly due to parasitism. Yellow membranes indicate jaundice which may be caused by liver disease. Jaundice is often accompanied by brown urine. Grey-blue membranes indicate poor oxygenation of the blood, which can be caused by exhaustion, severe pneumonia and hydrocyanide poisoning from young star grass. Dryness of the mucous membranes is a sign of dehydration, which is often associated with tight, inelastic skin; healthy skin is soft and elastic.

The **urine** also should be inspected routinely. Signs of blood breakdown products in the urine indicate diseases such as babesiosis and leptospirosis.

Animals should be inspected for abnormalities and ticks removed if present. **Feet** should be trimmed routinely. **Teats** often become sore because of milking and can be treated with creams to prevent irritation. Animals can be given medicine in tablet form, liquid form as a drench or by injection.

Restraining

In order to inspect an animal it might be necessary to restrain it and there are a number of ways of doing this. The easiest is to put the animal into a crush. Alternatively the animal can be thrown to the ground using ropes.

First aid

Prompt attention to small problems can prevent them becoming bigger problems. Cuts and wounds should be treated immediately to help them heal, reduce stress and stop secondary infection. Flies are attracted to wounds and spread infection. Sores firstly should be washed with boiled, soapy water and then one of several antiseptics can be applied.

Quarantine and acclimatisation

Good husbandry is of paramount importance to animals which are being introduced for the first time on to a farm or into a new area. Such animals should be kept in isolation for a period of time. This is called **quarantine** and allows animals to be screened for diseases so that they do not introduce disease to other stock. The period in quarantine depends on the incubation period of the most important diseases suspected.

This period also allows the animals to acclimatise to new conditions, new diets and new management routines. **Acclimatisation** is the adjustment or adaptation of their physiological and metabolic processes to new environmental conditions, including temperature, humidity, food, daily routines and disease organisms. Stock which have been moved will be nervous of new stock handlers and milkers, as well as of the new environment and need to be handled regularly and calmly to help overcome this.

Husbandry calendars

The husbandry of milk cows can be facilitated by working to a calendar of events which need attention at different times of the year. For example, vaccinations should be carried out at appropriate times of the year when they will not cause stress and will not interfere with peak lactation or calving.

There are numerous types of aids to the management of larger dairy herds which allow daily changes in the status of individual animals to be recorded quickly and easily on a visual display board. These are particularly concerned with calving and pregnancy status of cows.

Records

A further important aid to good husbandry is the keeping of records of all animals and events relating to animals throughout their lives. Records of inseminations, calving dates, health problems, treatment, milk yield and feeding schedules can help a farmer to predict future problems and focus attention on animals which need special care. It is best to have well organised records for each animal.

Some method of individual animal identification is necessary to make record keeping possible. Animals are best marked with ear tags, tatoos or brands.

Environmental control and housing

An important way to provide good conditions and to reduce stress for stock, is by reducing the adverse or extreme effects of climate. The animals' environment changes both seasonally and diurnally. Differences between wet and dry seasons are quite marked, as can be the differences between night and day.

All farmers practice some form of environmental control. Pastoralists take their cattle to better pasture, cut browse for their cattle, put them in a corral at night to protect them and light fires to keep away flies. Farmers in Kenya build *bomas* and farmers in Malawi build *kholas* for their cattle. The more intensive the system, the greater the degree of environmental control which farmers exert.

Fig 5.2   *Calf shelter with thatch roof*

Shelters

By reducing environmental stress, shelters and houses can improve milk production. Shelter can protect milk cows from excessive heat (and cold); if kept clean can reduce disease hazards; and can make management easier.

Roofs

The roof is the most important part of a shelter in a hot climate and even cheap structures can provide good protection if well designed (Fig 5.2). The temperature relationship between a cow and its environment is shown in Fig 5.3. By protecting animals from rain and by reducing long-wave heat loss at night, shelters can help to prevent cold stress.

The shadow of the roof should always cover the area in which the animals are confined. The temperature of the upper surface can become very high, up to 55°C higher than air temperature. The temperature of the lower surface should therefore be kept as low as possible which can

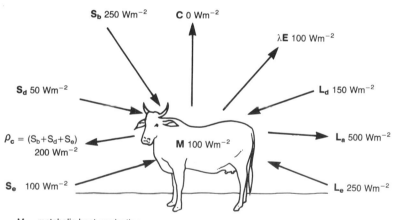

S_b 250 Wm^{-2} C 0 Wm^{-2}

λE 100 Wm^{-2}

S_d 50 Wm^{-2}

L_d 150 Wm^{-2}

$\rho_c = (S_b + S_d + S_e)$ 200 Wm^{-2}

M 100 Wm^{-2}

L_a 500 Wm^{-2}

S_e 100 Wm^{-2}

L_e 250 Wm^{-2}

M = metabolic heat production
S_b = incident short-wave radiation in direct beam
S_d = incident diffuse short-wave radiation from sky
S_e = short-wave radiation received after reflection from environment
$\rho_c = (S_b + S_d + S_e)$ = short-wave radiation reflected away from animal
L_d = long-wave radiation received from sky
L_e = long-wave radiation received from environment
L_a = long-wave radiation emitted by animal
C = heat loss by convection
λE = heat loss by evaporation

Fig 5.3 *The temperature relationship between a cow and its environment*

be achieved by using a good insulator so that the rate of heat transfer is low, increasing the reflectivity of the upper surface, sprinkling the roof with water, allowing an upward movement of air through the roof or making the roof high to increase the ventilation.

Good Extension Programmes would help inform farmers of the ways of making suitable roofs for stock houses and shelters. Stover stalks, thatch and palm fronds make good and cheap shade roofs.

Walls

Shelters can be fully walled, partially walled or open-sided to allow maximum ventilation. Air movement through the shelter keeps the inside air temperature similar to that outside by removing heat, moisture and pollutants. Most buildings rely on natural **convection** and the design must ensure that ventilation rate is adequate in all weather. Free convection is encouraged if walls are open, the roof has a steep pitch and the ridge is open or if the roof itself is permeable to air.

Some houses have thick, solid walls which have a large thermal capacity. These are designed to absorb rather than transmit the incident short-wave radiation during the day and release it again at night. Such houses are found in arid and semi-arid areas where there is a large diurnal temperature variation. They can be used for small numbers of animals since only low ventilation rates are possible to remove moisture and pollutants.

Floors

Floors have comparatively little effect on the thermal environment and their design depends largely on health and hygiene criteria. Unless the faeces and urine are disposed of properly, the environment becomes moist and ideal for the growth of pathogenic micro-organisms. Good drainage is essential. Dung can be dealt with by frequent removal, by allowing it to build up in bedding (deep litter) or by having a slatted floor. If possible, floors should be sloping and channels built-in to allow urine to run away. Houses should be disinfected regularly to reduce the numbers of flies and other parasites. A build up of infection and infestation sources can develop, for example during the calving season. Calf mortality at the beginning of the calving season can be as low as 2%, but may increase to 10% by the end of the season.

Site

The siting and orientation of a house affects the environment inside. Houses on exposed sites, such as higher ground or in clearings, benefit from better ventilation. Tall trees near to the house shade the roof and help to keep it cool.

Milk cows should be kept as near to the milking shed or homestead as possible, to minimise energy expenditure during walking. This is often not possible in extensively grazed systems, but attempts should be made to reduce the amount of walking that cows do each day. Stall-feeding with cut-and-carry fodder can help reduce the dependancy on distant grazing. Cows should have water available at all times, since they are lazy and will cut down on drinking rather than walk any distance, with drastic effects on milk yield.

Increasing the nutritive density of diets helps alleviate some heat stress problems. High concentrate plus low roughage diets have low heat increments (see p. 101) and therefore reduce heat produced inside the animal's body by metabolic and other activities. Stall-feeding reduces heat produced from activity. Grazing should be done at night – provided that the farm is secure and fenced. This practice allows an animal time to lose the heat built-up in its body during the day and to take advantage of cooler conditions for grazing when food intake will be greater. Also of note is the fact that sand floors cool down quickly at night and help heat loss.

Health care of milk cows

Good health care involves a number of routine practices. Important aspects of husbandry, hygiene, housing and climatic control have already been discussed and provide a basis of good health care. In addition the farmer must pay attention to nutrition, feeding and food hygiene and the prevention of infection with pathogenic organisms.

Some diseases can be passed from cattle to man. These are known as **zoonoses** and include anthrax, brucellosis, tuberculosis and ringworm.

General disorders and diseases

General problems that may arise with dairy animals include eye problems, digestive disorders, nutritional deficiencies (see chapter 6), problems at calving, sore teats, foreign bodies in the digestive tract, poisoning, snake bites, infertility (p. 79), mastitis, wounds, sprains, foot problems, lameness, skin problems and tick infestations.

Indigestion

Indigestion is caused by indigestible roughage, mouldy feed, excessive grain or concentrate intake and/or sudden changes in the diet. With experience, the farmer should be able to anticipate the onset of this condition and correct it accordingly.

Bloat

Bloat is usually caused by feeding too much, rich pasture (particularly legumes) which produces froth and gas in the rumen and can cause death within minutes if it is not treated. Froth and gas can be removed through a canula inserted into the rumen through the abdominal wall with a trocar. A stomach tube should be inserted if possible and the animal can be dosed with an anti-frothing agent. (Drenching with any vegetable oil or medicinal turpentine is better than no treatment at all.) The provision of dry fodder before cattle are put out to pasture in the morning and initially restricting the grazing time on rich pastures will help prevent the problem.

Metabolic disorders

Ketosis

Occurring in high yielding lactating cows a few days or weeks after calving, ketosis is caused by low blood glucose levels. The condition is best treated by intravenous administration of glucose.

Milk fever (hypocalcaemia)

Milk fever usually occurs within 48 hours after calving as a result of low blood calcium levels. It is characterised by an S-kink in the neck, muscular weakness and staggering, depressed consciousness and if untreated, coma and death. The condition is treated by an injection of 400–800 ml calcium borogluconate solution (25%).

The major cattle diseases are anthrax, black quarter, brucellosis, campylobacteriosis, coccidiosis, contageous bovine pleuropneumonia, diarrhoea, foot-and-mouth disease, footrot, haemorrhagic septicaemia, infectious bovine rhinotracheitis, leptospirosis, mange, mastitis, rinderpest, ringworm, salmonellosis, streptothricosis, trypanosomiasis and tuberculosis.

Ticks and tick-borne diseases

There are a number of important tick species which affect tropical cattle (see Table 5.1). The life-cycles differ depending on the species of tick and may involve one, two or three **hosts**. Ticks are found most commonly in the wet season on tall grasses and attach to grazing cattle. There are four stages in the life-cycles: **egg**, **larva**, **nymph** and **adult**. A female tick can lay 2000–20 000 eggs in the soil which hatch within 2–10 weeks depending on species and weather. Larvae climb on grass stems and leaves, attach themselves to a passing animal and subsequently feed on the blood of this acquired host. If the species is a one-host tick larvae stay on the animal (its host), and change into a nymph which feeds again and turns into an adult; the female adult feeds again, mates and then drops off to lay her eggs in the soil. In the life-cycle of two-host ticks, the larvae and nymphs stay on the same animal but the adult is found on a different animal. In three-host ticks, each stage (larvae, nymphs and adults) of the life-cycle attaches to a different host animal.

The different life-cycles of ticks complicate the effective use of control measures. The interval between dipping or spraying is determined by the life-cycle of the tick and the season.

Table 5.1 Main tick species affecting cattle

Species of ticks	Tick-borne diseases (TBD)
One-host ticks	
Boophilus decoloratus (blue tick)	Babesiosis, anaplasmosis
B. annulatus	Babesiosis
B. microplus (cattle tick)	Babesiosis, anaplasmosis, theileriosis
Two-host ticks	
Rhipicephalus evertsi (red legged tick)	East Coast fever (ECF), babesiosis
R. bursa	Babesiosis, anaplasmosis, theileriosis, rickettsiosis
Hyaloma truncatum (African bont legged tick)	Sweating sickness of calves
H. rufipes (large bont legged tick)	Abscesses
H. dromedariae (camel tick)	*Theileria annulata* infection
Three-host ticks	
⅄ *Rhipicephalus appendiculatus* (brown ear tick)	ECF (main vector of *T. parva*), corridor disease, babesiosis
R. simus, *R. pravus* and *R. capensis*	ECF (*R. simus* also transmits anaplasmosis)
✕ *Amblyoma hebraeum* (the bont tick)	Heartwater, tick-borne fever (*Rickettsia canorii*)

Tick control

One-host ticks (*Boophilus* spp.) have vulnerable larvae which will die from dessication or starvation within 1–3 months, depending on climate, if no hosts are available in the pasture. If pastures can be kept free from stock for about 6 weeks in the dry season and stock dipped or sprayed before return to these pastures, the stock will remain free of ticks. If only *Boophilus* ticks are present, it is possible to combine rotational grazing with strategic application of an **acaricide** (2–6 dippings or sprayings at 3 week intervals) carried out in the wet season. In the dry season, grazing around paddocks in rotation (i.e. rotational grazing) should keep the tick populations to a minimum.

For two- and three-host ticks a more rigid dipping or spraying regime is required with a weekly routine being the norm.

If dips are used, these require strict supervision to ensure that the concentration of acaricide is correct and does not become diluted or over polluted. It is also important to ensure that animals are fully submersed in the dip solution and that they enter and leave the dip properly to avoid injury.

In intensive systems with good fencing and control of stock movement, reduced dipping should be possible, since the spread of ticks occurs only by the movement of the host animal. If land is cleared of ticks by cultivation of crops prior to sowing pasture or burning, and cattle are dipped to remove ticks before being introduced to the pasture, then reduced dipping is feasible. Wild hosts, such as game animals, may also introduce ticks on to clean pasture.

If hand-spraying is used, care should be taken to spray correctly, starting on the backs, then proceeding downward and under, including genital organs, chest and brisket, testicles, base of the tail and inside the ears.

1st dip cattle before → pasture.

Tick-borne diseases

Ticks spread a number of diseases (Table 5.1), cause loss of blood, hypersensitive reactions at the biting point and secondary infection. They also cause a condition called **tick toxicosis** (tick paralysis), which results from toxins introduced to the animal by the tick. Sweating sickness in calves is a form of tick toxicosis.

Calves receive a degree of immunity to tick-borne diseases (TBD) from the dam. In areas where tick-borne diseases and their vectors occur, animals exposed to a sufficient number of infected ticks are infected early in their lives, suffer a relatively mild infection, usually recover and grow up resistant to further infection. If complete tick control cannot be achieved, calves need to be exposed to moderate numbers of ticks in order to develop resistance. If calves are brought up in tick free areas, they will be susceptible to tick-borne diseases and in adult life may suffer losses if control breaks down or animals are moved to other areas. This is a danger for nomadic cattle when they are moved into new areas.

Tick-borne diseases include anaplasmosis, babesiosis, East Coast fever and heartwater. The control of these tick-borne diseases is principally through the control of the tick vector. Vaccines have been developed and may be used to protect selected stock. Early cases can be successfully treated with appropriate drug regimes.

Body condition

A cow's body condition is a good indicator of her nutritional status and health. Body weight or heart girth measurements do not define the condition of the animal. Cows can often be allowed to lose body condition during the first part of lactation, but continued loss of condition will have serious repercussions for reconception, calf birth weight and the next lactation. In pastoral systems, cows usually have poorer condition than bulls of the same size or age, since the nutritional demands on a cow are greater than on the male animal and they can easily go into negative energy balance.

Condition scoring

Two systems are recommended for monitoring body condition: one for Zebu (Table 5.2) and one for taurine dairy cattle (Table 5.3). Both systems use a six-point score from 0 to 5 to indicate body condition. Low scores indicate poor condition and high scores good condition.

Zebu cattle

Zebu cattle are scored by first considering the second thigh muscle (Fig 5.4). The fat cover of this muscle is assessed from the side and cows can be divided into three groups on this basis:

scores 0 to 1 – muscles are wasted to give a concave appearance;

scores 2 and 3 – muscles have a straight appearance; and

scores 4 and 5 – muscles bulge and have a convex appearance.

Individual scoring is then made on levels of fat cover. In Zebu cattle, the fat cover of four areas is considered – on the spinous processes of the lumbar vertebrae, over the lower rib cage, at the hip bones (*tuber coxae*) and at the second thigh muscle.

Taurine cattle

The fat cover at the tail head is also considered.

Fig 5.4 *Second thigh muscle in a* Bos indicus *cow of condition 3*

Table 5.2 Body condition scoring for Zebu cattle

Score	Description
0	Emaciated animals with no apparent subcutaneous fat. Spinous processes in the lumbar region feel sharp.
1	Spinous processes are sharp, but less so than in score 0. There is some subcutaneous fat on these processes and on points of the hips.
2	Individual spinous processes are still fairly sharp to touch and ribs can be seen individually.
3	Spinous processes can be felt, but have a rounded feel and ribs cannot be seen individually.
4	Individual spinous processes can only be felt with firm pressure and points of hips are covered with fat and are rounded.
5	Spinous processes cannot be felt even with firm pressure. Animal has a blocky appearance.

Table 5.3 Body condition scoring for taurine cattle

Score	Description
0	Animal is emaciated with spinous processes, hip bones, tail head and ribs projected prominently. No fatty tissue can be detected; neural spines and transverse processes feel sharp.
1	Individual spinous processes are still fairly sharp to the touch and there is no fat around tail head. Hip bones, tail head and ribs are still prominent, but appear less obvious.
2	Spinous processes can be identified individually when touched, but feel rounded rather than sharp. There is some tissue cover around tail head, over hip bones and flank. Individual ribs are no longer visually obvious.
3	Spinous processes can only be felt with firm pressure. Areas on either side of tail head now have a degree of fat cover which can be easily felt.
4	Fat cover around tail head is evident as slight 'rounds', soft to the touch. Spinous processes cannot be felt even with firm pressure and folds of fat are beginning to develop over ribs and thighs of animal.
5	Bone structure is no longer noticeable and animal presents a blocky appearance. Tail head and hip bone are almost completely buried in fatty tissue and folds of fat are apparent over ribs and thighs. Spinous processes are completely covered by fat and animal's mobility is impaired by large amount of fat carried.

Lactation management

Since the costs of housing, health care and feeding are largely fixed, the more milk that can be produced, the lower the overall production cost. It is cheaper to produce 20 kg of milk from one cow than 10 kg each from two cows. Intensive systems aim to achieve the highest level of milk production, but dual-purpose systems achieve lower levels of economic production.

> *Whatever the production system, the objective of management is to achieve the best lifetime productivity of a cow by optimising production from each lactation.*

Season of calving

Under natural conditions, milk production may be seasonal and depends on nutrition and food supply which affect lactation and calving patterns. If breeding is not controlled and stock depend on natural food supplies, it is likely that a peak of calving will occur towards the end of the dry season and beginning of the wet season. In intensive systems, a seasonal pattern of production may be the easiest to achieve, but if there are financial incentives to produce milk throughout the year (and particularly in the dry season when prices may be higher), it may be preferable for the cows to calve at the beginning of the dry season when natural food supplies are poorest. If conserved forages, such as hay and silage, are available, dry season calving may be desirable.

Steaming-up

Many factors influence milk yield and the first concern is the condition and health of the cow in late pregnancy. If the cow is not suffering heat stress and is not ill, then nutrition determines whether the cow enters lactation in good or poor condition.

The cow must be fed in late pregnancy to meet the needs of the growing fetus and mammary gland development. Milk yield depends partly on the quantity of milk secretory tissue laid down in the mammary gland during the last month of pregnancy. The best lactations occur after a 60-day dry period with adequate feeding (steaming-up). In early lactation, dairy cows use body reserves to support milk synthesis: cows in good condition are better able to do this than cows in poor condition. Approximately 45 kg body fat will provide energy for 400 kg milk. The cow's condition in the early part of lactation also plays a part in ensuring that she will be able to conceive again at the right time.

69

If the season of calving is at the beginning of the wet season, food supplies for steaming-up may be in short supply. If conserved fodder is available in intensive systems or can be purchased by small-scale producers, this could be used for steaming-up at this period.

Daily management for optimum milk yield

Once the cow has calved (in addition to good husbandry practices), the farmer must pay attention to milking routine, feeding and health.

Milking routine

The mammary gland will continue to secrete milk at the maximum daily level if milk is removed regularly and as fully as possible at each milking. If milk is not removed, the rate of secretion will be reduced because as milk is secreted into the alveolus, the epithelial cells secrete against an increasing pressure. When this pressure becomes too great, the rate of secretion reduces. Secretion slows down and stops about 35 hours after the last milking.

The best time to milk is every 9–10 hours, but this would require more than twice a day milking. Milking three times a day produces 15–25% more milk than twice a day milking, but the labour cost and management inputs required are usually too great. In intensive systems an 11–13 hour

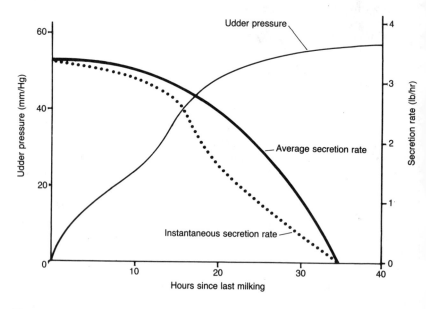

Fig 5.5   *Relationship between udder pressure and secretion rate*

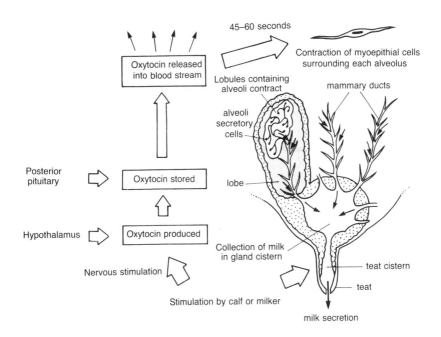

Fig 5.6 *Milk let-down response*

split between milkings is adopted. In pastoral systems cattle are often only milked in the morning, but the calf suckles for part of the rest of the 24 hour period, thus maintaining milk secretion.

Milk let-down is controlled by the hormone oxytocin which is released as a reflex to natural stimuli (Fig 5.6). The usual stimulus is the calf beginning to suckle, but routine activities such as the attendant coming into the milking parlour also act as a stimulus. If milking is inefficient and slow, all the milk will not be let down and milked out. Normally, there is usually 15% of the total milk volume remaining in the udder after let-down. Incomplete milking reduces the long-term yield performance.

The method of milking (by machine or hand) also affects the amount of milk produced. Restricted suckled cows have been observed to produce more milk than cows which are not suckled.

Short-term stress due to noise, excitement and other disturbances can affect milk let-down, since adrenalin secretion inhibits milk let-down. High environmental temperatures cause stress. At ambient temperatures over 24°C, milk yields may be affected due to lower food intake, temperature control activity and behavioural responses, such as shade seeking. This can be compensated for by adjusting feeding to a low dry matter concentrate, time of feeding, night feeding and by the provision of shade and the orientation of housing.

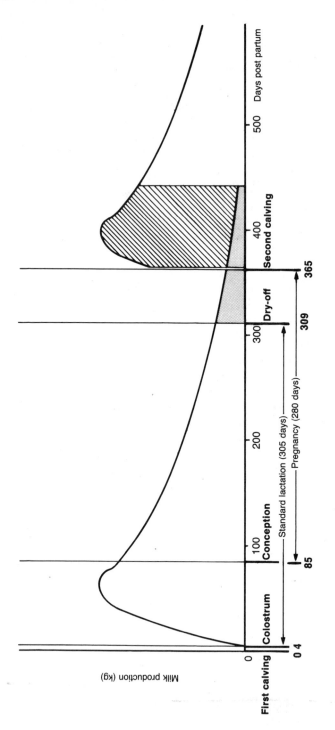

The loss of production resulting from drying-off [ ] is more than compensated for by the additional milk produced at the beginning of the next lactation

Fig 5.7 *The 305-day lactation and the ideal cow year*

Maintenance of health

> *A sure sign that the animal is suffering from an infection is a drop in milk yield.*

Any form of acute infection will affect milk yield, but subclinical disorders, such as mastitis, can also reduce yield. Metabolic disorders, such as milk fever and ketosis, can both affect milk production in early lactation.

Feeding routine

It is important to maintain a routine feeding pattern. Cows should not be fed irregularly and milking time should be fixed in relation to feeding. Disturbances in feeding and milking routines can cause a drop in milk yield.

The 305-day lactation

Once the cow has calved the farmer must:

- provide suitable conditions for the cows to produce milk from the present lactation
- prepare for milk production in the next lactation.

This requires good day to day husbandry and attention to feeding, health and milking routines and that the cows conceive again at the right time. The 'right time' depends on the anticipated calving interval (the period between consecutive calvings). Maximum life-time production is achieved with a calving interval as near to 365 days as possible. This is based on a 305-day lactation and a 60-day dry period prior to calving (Fig. 5.7). To achieve this, a cow must conceive before 85 days have elapsed after calving. Much of dairy cow management is aimed at achieving this goal, of which management of fertility is an important component. In practice, calving intervals are often longer than 365 days. In Malawi, for example, mean calving intervals for crossbred cows (Local/Friesian) of 485 days and lactation length of up to 392 days have been reported. If natural service is not used, good oestrus detection and efficient AI are essential to ensure that cows conceive.

The characteristics of the lactation curve of high yielding dairy breeds are such that the additional milk produced by starting a new lactation compensates for the milk that could have been produced by extending the present lactation. Figure 5.8 shows the typical lactation curves for Holstein/Friesian, Friesian × Bunaji and pure Bunaji cows at Shika Agricultural Research Station, Zaria (1950–78).

The milk lost in the shaded area of the first lactation in Fig 5.7 is more than compensated for by that produced in the next lactation in the cross-hatch shaded area. An example of the quantitative effect on overall milk yields (kg) of the length of the calving to conception interval for Friesian cattle in a country such as Britain is shown in Table 5.4.

These reductions are for individual cows, but if they are multiplied by the whole herd of say 50 cows, they amount to a substantial loss of milk and revenue. The calving-to-conception interval may be critical when considering the economic feasibility of dairy enterprises.

Table 5.4 Effect of calving-to-conception interval on milk yield

| Post-calving interval-to-conception (days) | Average milk yield (kg) | | | | | |
| | Lactation number | | | | | |
	1	2	3	4	Total	Difference
<96	3885	4120	4981	4969	17955	–
96–125	3679	4095	4969	4725	17468	487
>125	3671	3877	4717	4406	16670	798

Source: Russell (1980)

Management of reproduction and fertility

Infertility causes economic loss in both extensive and intensive milk producing enterprises. The causes are many and often not immediately apparent. An understanding of the reproductive cycle helps to anticipate problems which might arise and to provide solutions. The husbandry practices (see earlier) are the first step towards achieving good fertility and optimum reproductive rate. Good fertility is essential to maintain high levels of milk production and the financial viability of intensive systems.

The reproductive tract

Ova are formed in the ovaries situated in the upper part of the abdomen and pass into the oviducts (Fallopian tubes) before passing into the uterus. In the cow, the uterus has two horns which connect with the main body of the uterus. The entrance to the uterus is the cervix, which joins the vagina and finally the vulva which is the external entrance to the reproductive tract (Fig 5.9).

Spermatozoa, the male germ cells, are produced in the seminiferous tubules of the testes (testicles) located in the scrotum (Fig 5.10). Spermatogenesis requires temperatures just *below* normal body temperature, which is why the testicles are found outside the body cavity.

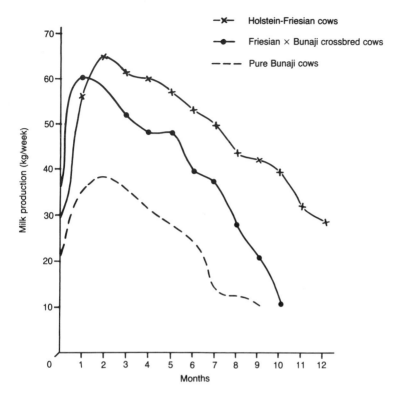

Fig 5.8   *Lactation curves of Holstein/Friesian and Friesian × Bunaji cattle*

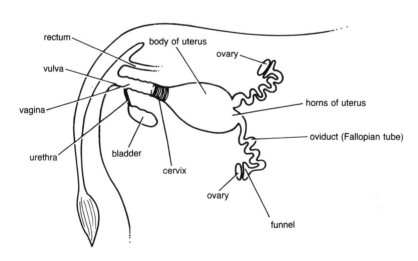

Fig 5.9   *Reproductive tract of the cow*

75

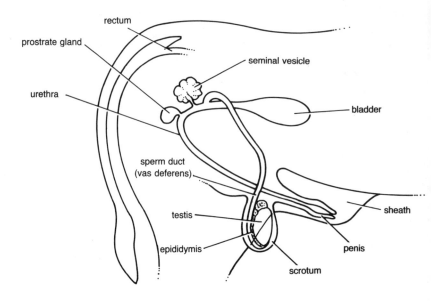

Fig 5.10 *Reproductive tract of the bull*

Optimum fertility and indices of fertility

There are a number of indices of fertility based on different criteria and definitions. The best measure of an animals fertility is its ability to produce a second generation of progeny. Usual indices accept more limited definitions of fertility and include **weaning percentage** (the number of calves weaned out of the number of cows put to the bull or AI), **calving rate** or **calving percentage** (the number of calves born out of the number of cows put to the bull or AI), **calving interval** (the number of days between two consecutive calvings), **conception rate** (the number of cows conceiving out of the total put to the bull or AI), **inseminations per conception** and **calving to conception interval** or **days open** (the number of days from calving to the next conception). These are all measures of the 'success' of reproduction.

If parturition is taken as the beginning of the reproductive cycle in a cow, then **optimum fertility** occurs when a cow

1 recommences cycling soon after calving (within 42 days),
2 conceives within 85 days,
3 produces a viable zygote which implants in the uterus and
4 survives to full term to produce a healthy calf which survives to adult-hood to continue the reproductive cycle.

76

The reproductive cycle can be divided into a number of stages at which problems might arise. These are parturition, onset of the oestrous cycle, conception (fertilisation), the zygote (to day 6), embryo (day 6 to implantation on day 42–45), the fetus and parturition again.

Gestation varies between 274 and 291 days with a mean of 280 days. After parturition, attention must be given to a number of factors to ensure timely reconception. The cow should be in good body condition; the uterus must return to its non-pregnant state of size and health (i.e. it must involute); and the cow must cleanse properly and expel all parts of the afterbirth. Sometimes, even in cows which calve normally, this does not occur and infection may set in. Problems are more likely to arise if the cow aborts or if it is suffering from an infection of the uterus or genital tract. In a healthy cow involution should occur within 28 days.

Oestrous cycle

Once a cow has calved, the next important stage is the onset of the oestrous cycle and the showing of **heat** or **oestrus**. The reproductive cycle is controlled largely by the hormones of reproduction, but can be influenced by nutrition, environmental factors and aspects of health.

The main endocrine glands associated with reproduction are the hypothalamus and the pituitary gland. The main reproductive hormones are follicle stimulating hormone (**FSH**) and luteinising hormone (**LH**). At the point where the egg is released from the ovary, a corpus luteum forms which produces **progesterone**. This hormone stimulates the uterus in preparation for the egg. It is also responsible for maintaining pregnancy and inhibiting further ovulation during pregnancy. If fertilisation does not occur, the corpus luteum regresses and ovulation occurs again.

Ideally the cow should start cycling within 42 days after calving. The onset of the cycle can be delayed because of poor nutrition, retained placenta, metritis and delayed involution of the uterus. The overall cycle length is between 18 to 24 days.

If natural service is not used, the key to a short calving to conception interval is good **oestrus detection**. In the cow, the period of oestrus is on average 18 hours. Efficient recording of signs and events can help detection.

Main signs of oestrus (a)   when a bull is used:
    bellowing
    the cow being mounted
    smelling other cows
    restlessness
    swelling of the vulva
    mucous discharge.

Fig 5.11 *Cow standing to be mounted*

Main methods of oestrus detection (b) when a bull is not used:
watching for cows standing to be mounted
watching for vaginal discharge
watching for abnormal signs of behaviour
using tail paint
consulting records.

Fertilisation

If an egg is shed from the ovary, fertilisation depends firstly on the insemination of viable spermatozoa from a fertile bull with good **libido** (willingness and ability of the bull to mate). If AI is used, this must be carried out at the right time, which means that the farmer should watch the cows to detect oestrus.

Fertilisation usually occurs in the upper third of the **oviduct** (Fallopian tube) and within six hours of ovulation. If fertilisation is successful, then a **zygote** (fertilised ovum) is formed which begins to pass down the oviduct. During passage through the oviduct the cells begin to double and redouble so that by the time the ovum reaches the uterus it has about 32 cells. The rate of progress of the ovum through the oviduct should be such that it arrives when the uterine environment is compatible with further development. At the 16- to 32-cell stage when it enters the uterus, the embryo is known as a **morula**. This develops into a **blastocyst** which elongates to fill most of the lumen of the uterus.

Survival of embryo

The major proportion of embryonic loss occurs before day 15 after service. Implantation in cattle occurs at approximately day 42. Once an embryo

78

has implanted, losses may occur because of stress arising from environmental, nutritional or infectious causes. Losses arising from environmental stress should be uncommon and losses at the embryonic stage are more likely to be the result of nutritional and congenital problems. Embryonic death and reabsorption of the embryo will occur if the animal's nutritional status declines.

Survival of fetus

Fetal losses are usually associated with infection rather than with nutritional deficiencies. Abortion can occur because of an infection with a number of disease organisms, which include campylobacteriosis, leptospirosis, brucellosis, Rift Valley fever, infectious bovine rhinotracheitis (IBR), mucosal disease and trichomoniasis. Cattle can be vaccinated against the first five of these. Campylobacteriosis, trichomoniasis and IBR can be controlled by using AI if bulls are infected. Often these diseases are introduced when other stock come into contact with the herd and some degree of control can be achieved by keeping as closed a herd as possible. Range cattle might also come into contact with aborted fetuses and fetal membranes from which they could gain infection. A number of bacterial and fungal infections of the uterus also cause loss of the fetus.

Causes of infertility

Hormonal causes of infertility

Cystic ovaries result from hormonal imbalance and are most common in high yielding cows. They can cause irregular oestrus, continuous oestrus or anoestrus depending on the type of cyst.

Retained corpus luteum persists on the ovary as a result of infection of the uterus following abortion or early fetal death and the cow does not come into heat.

Silent heat allows ovulation to occur but there is little or no sign of heat. This occurs mainly because of poor nutrition.

Inactive ovaries (or anoestrus) is the result of poor nutrition, poor condition or sickness. Such cows may have inactive ovaries and do not come into heat.

Nutritional causes of infertility
Nutritional causes include under feeding (which causes inactive ovaries),

overfeeding, Vitamin A deficiency (also causing difficult calvings, abortion, retained afterbirth and infertility) and Vitamin E/selenium deficiency (causing retained afterbirth and infertility).

> *The predominant cause of long calving intervals in the tropics is loss of condition caused by poor nutrition and disease, which causes anoestrus, especially in seasonal rainfall areas.*

One of the most important practical points of management in seasonal rainfall areas is for the postpartum cow to become pregnant again before grass growth falls off in quality and quantity in the dry season. If reconception does not occur, the cow is likely to become anoestrus and not to cycle again until the new grass growth with the following rains. This results in intervals between calving of 2 years.

Attention to the bull

Bulls should be in good health when they are expected to serve the herd. No bull should be overworked. If a bull has been ill it should be rested and allowed to fully recover. Arthritis of the spine or hips may cause pain and stop the bull from working (mounting the cow), as may painful hoof conditions. High ambient temperatures reduce libido – the bull should be provided with shade.

Fertility records

Farmers should keep good records of events in each cow's life to assist in cow management. Important dates should be recorded, such as calving and service dates, as well as treatments for ill health. The farmer should keep a notebook to record these events and transfer them to a permanent Record Book. A blackboard in the milking parlour is a common feature in many intensive systems. More sophisticated Individual Cow Records can be kept and the most advanced dairy farms might have computerised recording systems.

6 Feeding milk cows

A more detailed account about feeding milk cows is given in *Ruminant Nutrition* by John Chesworth in this series *The Tropical Agriculturalist*.

Systems of feeding

Extensive grazing on communally owned land

Herding around villages by settled cattle owners and herding over longer distances by pastoralists are the two main systems of extensive grazing on communally owned land. Cows graze extensively on such land and receive a diet of good quality natural pasture in the early wet season, but a poorer quality diet later in the season. Grazing patterns are partly determined by the availability and distance of water from the grazing areas or village.

In communal land use systems, livestock owners cannot improve the pasture or water supplies and cannot improve or install infrastructure, such as fences. Forage conservation is not possible, though if cattle owners cooperated, pasture could be protected as a dry season grazing reserve of standing hay, as was the traditional practice. Where competition for land exists between cattle and crop producers, grazing land is usually taken over by crop producers. Conflicts may arise between livestock owners and crop farmers. Pastoralists, such as the Fulani in West Africa, are affected by such conflicts and grazing land is continually lost to crop farming, even in areas marginal for crop production. For these reasons it is difficult to plan better feeding regimes for milk producing cows based on grazing. Communally managed legume fodder banks (p. 92) have been developed in Nigeria and offer a potential solution to the problem.

Dry season grazing may be supplemented by crop residues, browse and regrowth of burnt areas and flood plains. In cultivated areas, providing food for cattle is more difficult, since browse is usually scarce, crop residues are likely to be finished by the mid dry season and there are few areas of riverine vegetation for cattle. Under these conditions cattle in the dry season usually lose weight, with consequent effects on fertility and future milk production. Cattle owners often have agreements with cultivators who allow cattle to graze grain stovers in return for the cattle dung deposited on their land. Supplementation and cut-and-carry inputs may be possible. Some cattle ·owners may be able to buy conserved fodder locally or from other areas and some suitable supplements may be available. If the cattle owners are settled, they may grow crops themselves and can use their own crop residues and by-products. Fodder trees and other fodder crops can be grown on small plots around homesteads on land not suitable for or not used for other crops.

Cut-and-carry systems

Where cattle owners are settled and where pressure on land is increasing, extensive grazing systems can be modified to include cut-and-carry fodder for cattle. A good example of this are the housed milking cows (usually Jerseys) kept on the slopes of Kilimanjaro, Tanzania – a method similar to larger-scale zero-grazing systems. Such fodder can be cut from river banks or other areas of more abundant green vegetation. Browse (tree leaves) also may be cut for feeding cattle. However, cutting fodder involves an increased labour input to the cattle enterprise.

In areas where farmers cannot fence land, it might be better for milk producing animals to be stall-fed from improved grass/legume pasture grown on fallow land. This also would reduce the energy loss from walking and the stress of grazing in the heat of the day and improve milk

Fig 6.1 *Milk cow (local East Africa Zebu) fed cut-and-carry fodder, in Kenya*

Fig 6.2 *Napier grass (*Pennisetum purpureum*) growing beside a path in a village, in Ethiopia*

yields as well as avoiding conflicts with crop farmers. In Kenya, farmers plant grasses, such as *Panicum coloratum* var. *makarikariensis* and Bana grass (*Napier* spp.), on their farms as aids to soil conservation along contour furrows (*fanya juu*) and may use this for their stock. Similarly, in Ethiopia, Napier grass can be grown successfully in back-yard plots. Cut-and-carry inputs may act as a supplement to natural grazing and it may be possible for farmers to plant small plots of improved fodder varieties, such as Napier grass, or fodder trees such as leucaena. In Central America, sugar cane is used for this purpose.

Fig 6.3 *Friesian cattle grazing sown pasture, in Malaysia*

In more intensive larger-scale systems, cows may be zero-grazed for all or part of the year with bought-in food fed in the stall or yard. This avoids the requirement to have grazing land near to the farm and is the most common system for large, intensive units in Libya, the Middle East, Saudi Arabia and the Gulf States. Diets may include a larger portion of concentrate foods as well as forage and by-product foods which can be obtained locally and transported in bulk to the farm.

Small-scale semi-intensive systems

Intensive feeding systems in temperate regions utilise a high proportion of grain-based concentrates, but these are not available as livestock foods in many tropical countries. Semi-intensive production requires additional inputs if higher yields are to be achieved. Milk schemes often recommend the utilisation of improved pasture and supplementation with locally available concentrates. Concentrates are expensive, of unpredictable quality and often not available because of transport and supply problems. More reliable feeding systems for small-scale farmers need to be developed, based on home-grown foods, such as Napier grass and leucaena, grain millings, urea and locally available concentrates. Better utilisation can be made of high yielding tropical crops, for example sugar cane which has been recommended in Latin America to produce moderate levels of milk at lower cost. The use of such crops would be feasible if the price received for milk was enough to cover the costs.

Intensive grazing systems on sown pastures

Sown pastures may be grown on larger farms, but less than 5% of tropical

Fig 6.4 *Natural pasture in the derived savanna zone of Nigeria*

pastures are estimated to be sown pastures. The sowing of improved pasture is possible where land is owned and if milk production can compete with grain crop cultivation as a form of land use.

Foods for cows
Roughages

Most tropical cattle depend on natural pasture, crop residues and crop by-products. Except for the early growth stages, unimproved tropical grasses are poor quality for most of the year. Despite this, in most feeding systems cows are expected to derive their maintenance requirement and the first few kilograms of milk from roughage. This is appropriate, since roughage is cheap and the cow requires a certain amount of roughage for proper rumen function. In more intensive systems supplementary concentrates form a greater proportion of the diet. Even in intensive systems, approximately 30% of the dry matter intake should be from roughage. The roughage can be natural or sown pasture, conserved forage such as silage or hay, forage from fodder banks, crop residues or browse.

Pasture

The management of natural pasture is difficult, since the aim of most livestock owners in areas where land is communally owned is to maximise production for each animal and not to optimise production for each unit of land. Animal production often occurs at the expense of the pasture, which results in overgrazing.

The natural pasture species of an area are usually well documented.

In Africa for example, the species of 47 countries have been grouped into 16 grassland types and 100 sub-types. Even though there may be good growth of natural pasture, cattle often cannot eat enough to meet the requirements for high levels of milk production and growth because of the low quality of the grass. Natural grasses mature quickly and soon become unpalatable and of low nutritive value. Young pasture is more nutritious than old pasture, but since land is often not owned, the grazier cannot manage pasture to keep it young and achieve the best pasture production.

Natural pasture can be improved by reducing tree cover and by over-sowing with legumes, such as stylo (*Stylosanthes guianensis, S. humilis, S. hamata*) and siratro (*Macroptilium atropurpureum*). Controlled grazing and controlled stocking rates help to preserve natural pasture, but community action would usually be required to achieve this. Levels of milk production are not high in animals which depend on natural pasture, unless supplementary feeds are given. Seasonal milk production occurs depending on the growth patterns of the pasture.

Pasture grasses and legumes

> *Grass and legume species differ in their suitability for different areas and countries depending on rainfall and mean seasonal temperature.*

Perennial grasses and legumes have the potential for year-round growth and production. The main influencing factors are energy supply (sunlight), carbon dioxide concentration, nutrient availability (mineral salts), water and temperature. Light is the primary factor, but photosynthesis can be reduced by temperature, water stress and soil nutrient supply. Water and nutrient supply can be modified by management according to the level of capital inputs available and the economics of grassland production compared with crop production.

Tropical grass and legume species can exhibit a wide range of adaptability, grass species more so than legumes. It is not unusual to find that species thrive from lowland equatorial regions across several latitudes and maybe beyond the tropics. Couch grass (*Cynodon dactylon*), jaragua grass (*Hyparrhenia rufa*), and Gamba grass (*Andropogon gayanus*) prosper in the forest zone of West Africa, but also extend across the Guinea and Sudan savanna zones. Among the legumes, glycine (*Glycine wightii*), stylo (*Stylosanthes guianensis*), siratro (*Macroptilium atropurpureum*) and leucaena (*Leucaena leucocephala*) show broad adaptation.

Tropical grasses differ from temperate grasses in that the former usually have lower crude protein and higher crude fibre levels when cut at similar stages of growth. Tropical grasses usually have higher dry matter contents. Nutritive value is influenced by the leaf to stem ratio, by the stage of growth at cutting or grazing, fertiliser or manure applications and climate.

Grasses are members of the family **Gramineae** and as well as pasture species grown for stock, include species grown for their grain, such as maize, millet, sorghum and wheat. It is estimated that there are 10 000 species of grasses, but only 40 or so of these are used for sown pastures. Grasses are either **annual** (completing their life-cycle in one year) or **perennial** (which are of two or more years duration).

A number of improved grass and legume species have been developed over the past few decades. Grasses include buffle grass (*Cenchrus ciliaris*), Columbus grass (*Sorghum almum*), Guinea grass (*Panicum* spp.), gamba grass (*Andropogon gayanus*), Kikuyu grass (*Pennisetum cladestinum*), love grass (*Eragrostis curvula*), molasses grass (*Melinis minutiflora*), Napier grass (*Pennisetum purpureum*), pangola grass (*Digitaria decumbens*), para grass (*Brachiaria mutica*), paspalum (*Paspalum dilatatum*), Rhodes grass (*Chloris gayana*), Surinam grass (*Brachiaria decumbens*) and star grass (*Cynodon* spp.).

Legumes are members of the **Papilionaceae** family, a division of the **Leguminosae** that includes over 10 000 species. They are able to convert atmospheric nitrogen into plant proteins through the action of bacteria of the *Rhizobium* genus that grow and multiply in nodules on the plant's roots. All legumes bear pods and include the bean and pulse varieties used for human food.

Improved legumes include centro (*Centrosema pubescens*), green and silver leaf desmodium (*Desmodium intortum* and *D. uncinatum*), Kenya white clover (*Trifolium semipilosum*), lablab (*Lablab purpureus*), leucaena (*Leucaena leucocephala*), lotononis (*Lotononis bainesii*), lucerne (*Medicago sativa*), puero (*Pueraria phaseoloides*), siratro (*Macroptilium atropurpureum*), stylo (*Stylosanthes guianensis*) and Townsville stylo (*Stylosanthes humilis*).

Intensive sown pasture

Intensive pasture management is usually the only way to provide high producing cows with large amounts of good quality pasture. There are few tropical farming systems in which this could be achieved, because land can usually be better used for crop production.

Pastures can be dryland, irrigated, fertilised, grass- or legume-based or a combination of these. Grass and legume species must be chosen for their ability to thrive together. Some species will respond better than others to extra water, some better to fertiliser and some are better than others for grass/legume mixtures. In Zimbabwe, the most suitable pastures are fertilised dryland pastures (for example star grass, giant Rhodes grass/siratro and star grass/silver leaf desmodium), heavily fertilised irrigated pastures (for example Kikuyu grass, star grass and Paraguay paspalum) or irrigated mixed pastures (Kikuyu grass/Kenya white clover). Improved pastures such as these are recommended for farmers with the resources to maintain them. In Kenya, planted leys of Rhodes grass, Nandi setaria (*Setaria anceps*) and Guinea grass are grown.

Pasture management

> *The aim of pasture management is to achieve a stocking rate to utilise the pasture without causing overgrazing or accumulation of mature grass.*

Some research has shown that if the same area of pasture is used constantly rather than dividing it up, the overall production of dairy cows can be higher. Rotational grazing systems are usually adopted to achieve optimum utilisation, using a system of six or eight **paddocks**, each grazed for a period of 8–10 days with a rest time for each paddock of 40–50 days; this keeps the pasture young and leafy. If pasture cannot be grazed, it can be cut after 6–7 weeks of growth and the grass conserved as silage or hay for use later in the season.

Stocking rate and productivity
Stocking rate is the number of hectares allocated to each livestock unit. The optimum stocking rate is difficult to define and may vary from season to season and year to year. **Carrying capacity** is the number of hectares of a vegetation type which will support one livestock unit for a full year and should not be confused with stocking rate.

As stocking rate increases, there is an initial increase in animal performance due to an associated decline in dead herbage which is of low nutritional value. Productivity per animal begins to decline as the selectivity of animals increases and the more palatable grass species diminish. Productivity per hectare also increases with increased stocking rate and begins to decline at a higher stocking rate than that which causes a decline of animal performance (Fig 6.5).

Grazing period
The grazing period for a paddock or open pasture should be such that optimum utilisation of the vegetation occurs. The grazing period must be balanced with the stocking rate so that the most palatable species are not overgrazed (because the grazing period is too long) or the less palatable species become too dominant (because the grazing period is too short). With a paddock system, the grazing period depends on the number of paddocks available and should allow each paddock to be grazed once

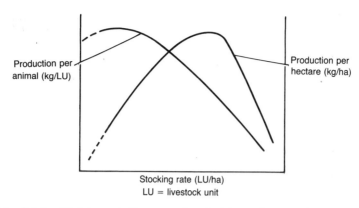

Fig 6.5 *Relationship between stocking rate and productivity per hectare and per animal*

each 60 days (in areas with over 700 mm/year rainfall). Thus if there are four paddocks, they should each be grazed for 15 days in rotation when pasture is in good condition in the mid-wet season. In the early growing season, the grazing period should be shorter to prevent overgrazing. The number of paddocks per herd should allow an optimum rotation. The optimum lies between four and eight paddocks.

Rest period

The rest period for pastures is as important as the grazing period. Rest during the rainy season permits leaf, plant and seed growth. In higher rainfall areas (over 700 mm/year), a rest period of 45 days is required, but in drier areas longer rest periods are required.

Removal of dead herbage

Old grass and tufts reduce the vigour of the growth by shading lower, younger growth. If the stocking rate, grazing period and rest period are correct then there should be little accumulation, except in years of above average rainfall. To avoid accumulation of old grass, different types of pasture should be made into separate paddocks. In addition, larger numbers of cattle can be concentrated in small areas for short periods at the end of the dry season to trample dead herbage. Fire can also be used to remove excess herbage.

These principles can be incorporated into one of three types of pasture management system that allow the greatest flexibility (stocking rate, grazing, and resting periods) according to the needs of the pasture and animals.

1 Continuous grazing is the simplest method and involves stocking a fixed number of cattle on the same pasture all season. The stocking rate should be well within the carrying capacity.

Fig 6.6.   Above left   *Silver leaf and green leaf desmodium (*Desmodium uncinatum *and* D. intortum   *on a small farm, in Ethiopia*

Fig 6.7. Above right   *Forage grass strips in a field of barley, in Ethiopia*

Fig 6.8   Sesbania sesban, *a legume, grown with sorghum, in Ethiopia*

2 Rotational resting system uses paddocks that are rested for long periods in the grazing season. The season can be divided into two or three periods and the paddocks rested for the whole of one of the periods.

3 Rotational grazing system with four to eight paddocks, allows animals move in rotation through each paddock in turn.

Backyard fodder production from sown species

Improved pasture grass and legume species are not only useful for large-scale sown pastures, but also for planting on small plots of land around farms and dwellings. These are also used for intercropping, undersowing or as forage strips in field crops. Grasses, such as Napier grass, and fodder legumes are planted around farm boundaries or along paths. Fodder trees are planted on uncultivable land, around dwellings or inter-cropped with food crops.

The total dry matter production by these combined methods, even on a small farm, could provide a milk cow with sufficient supplementary good quality fodder to support milk production throughout the year.

Conserved forages

Hay

Hay is high quality grass (early flowering stage) preserved by cutting and reducing the moisture content to less than 25%. Haymaking is more difficult in the wet and semi-humid tropics than in temperate regions, because of the time required to dry grass in the field. The best method is to allow grass to grow for the last 6–8 weeks of the rainy season and then to cut for hay when the rains are finishing or finished. This overcomes the drying problems. Grass can be sun dried, but if this is not possible it can be artificially dried in barns. Uncut grass left to dry in the field (standing hay) has lower food value than cut hay, because the former is usually more mature.

Silage

Silage is high quality grass preserved by allowing an anaerobic fermentation at a pH less than 4.0 in which soluble carbohydrates in the grass are fermented by bacteria to lactic acid. Lactic acid can be used by rumen microbes as an energy source and good silage has a high nutritive value for the ruminant animal. Fibrous grasses do not make good silage because of their low soluble sugar content and the difficulty of achieving anaerobic conditions in the silage clamp. The grass should be chopped and a form of soluble carbohydrate, such as molasses, added to speed up the fermentation.

Fodder banks

In Nigeria, the concept of the fodder bank has been pioneered by the International Livestock Centre for Africa. Fulani cattle in Nigeria produce little over 1 kg/day of milk at the height of the wet season in June. The grasses, even of sub-humid areas, will not support milk production for six or more months of the year due to low digestibility, intake and protein content. Grazing must therefore be supplemented if milk production above 1–2 kg/day is to be sustained.

To construct a fodder bank, it is recommended to use 4 ha land for 50 animals in the dry season. The land needs to be cleared and fenced. The seedbed is prepared by keeping cattle in the enclosure overnight at the end of the dry season. This ensures good manuring. A mixture of Stylo (*Stylosanthes guianensis* var. 'Cook') (6 kg/ha) and super phosphate (150 kg/ha) is broadcast by hand. Sowing should be carried out early in the wet season and repeated each year. Seeds should be allowed to germinate and then be grazed 6 weeks later to remove vigorous grass species. Once the legume has become dominant, grazing should be discontinued until the dry season. In the dry season cattle can be allowed to graze for 3 hours/day during which time the dry matter intake of stylo would be about 4 kg. The stylo should be grazed down to a height of 18 cm. Grazing should be ceased at the time of seed setting to allow seeds to be collected for the following year.

If the legume is well maintained, the cost of producing and maintaining a fodder bank should go down each year. Preference for use of the fodder bank should be given to young stock and milking cows, although in practice owners often cannot do this because of the complex multi-ownership of cattle.

Crop by-products and treated roughages

Cereal straws (e.g. sorghum stover, rice straw and to some extent maize stover) are important foods where they are grown in the tropics. They have high fibre and low crude protein contents and digestibilities are low (40% or less). If these foods are fed with a supplement, animals perform adequately. Poor quality roughages such as these can be treated with alkali (e.g. ammonium hydroxide – produced from ammonia or urea or sodium hydroxide). Treatment increases digestibility to 55% and if ammonium hydroxide is used, the crude protein content is also improved.

Trees, leaves and browse

On small farms and in drier pastoral areas, tree leaves provide an important source of green fodder as a supplement to other roughage foods and

Fig 6.9   Acacia albida, *(*left*) tree, (*right*) pods, in Ghana*

cereal straws. Trees can be planted around farms or on non-cultivable land.

The integration of trees into agricultural enterprises has received much attention, since trees also help to conserve soil and soil fertility, provide firewood, timber, poles, shelter, shade and a source of animal food. Farmers usually know the trees which are of value for different purposes and such trees are often left in the farm land. *Acacia albida* (Fig 6.9) is an example of a tree used for animal food. Its foliage grows in the dry season and pods are formed which are fed to livestock.

Cultivated fodder tree species include *Leucaena leucocephala* (Fig 6.10) and *Gliricidia sepium*. The latter can be used as a living fence. Leucaena can be grazed or cut and fed as a supplement to natural pasture or poor quality roughage. In Kano State, Nigeria *Albizia lebbek*, *Leucaena leucocephala*, *Dalbergia sissoo*, various *Ficus* species and baobab (*Adansonia digitata*) are recommended browse species and there are similar examples for other countries.

Concentrates, supplements and other foods

Other foods which are available for ruminants include concentrates, grain millings, industrial by-products (e.g. brewers' grains, molasses, urea and meals of blood, bone and fish) and crops (e.g. bananas and sweet potato vines).

Concentrates

Maize and sorghum meal are concentrates with high energy contents, whereas others, such as cottonseed meal, soyabean meal, palm kernal

Fig 6.10 *Leucaena* (Leucaena leucocephala)

cake, groundnut cake and sunflower cake, are good sources of protein. Smaller quantities of foods such as meat and bone meal, fish meal and blood meal may also be given to supplement nitrogen and protein intake. These nutrients might be given as pelleted concentrates made from various ingredients. In Nigeria, a two tier system of supplementation is recommended in combination with fodder banks, using agro-industrial by-products such as cottonseed cake and molasses, with urea and salt lick blocks.

Molasses

Molasses is a source of readily fermentable energy and is used as a carrier for urea. The mix is molasses (42 litres) plus urea (2 kg) in water (8 litres). When fed in a roller drum intake is reduced and such a drum could be used for a group of twenty or more animals (Fig 6.11). Alternatively, a mix of molasses (4 litres) plus urea (200 g) in water (90 litres) can be fed to two animals for a day (Fig 6.12). Also molasses can be mixed with chaff as a carrier and fed in a trough.

Urea

Not protein, but a simple compound (formula $CO(NH_2)_2$), urea is a valuable source of nitrogen. Pure urea has 46.7% nitrogen. The crude protein value of urea is the nitrogen content × 6.25 and 1 kg urea has a crude protein equivalent of about 2.92 kg. Urea used for animal feeds may have a slightly lower nitrogen content than pure urea and account should be taken of this when calculating the food value. Rumen microbes can use urea as a source of nitrogen and use it optimally if a source of readily fermentable energy such as molasses is available.

Minerals and vitamins

Mineral licks can be provided to ensure that stock receive a balanced mineral intake. Natural sources of salt and minerals are usually known to stock owners, but while these provide some of the mineral requirements of cattle, often they do not provide a balanced input of essential minerals. They usually contain sodium and iron, but are often low in other minerals. Salt licks, soil licks and well water all provide minerals.

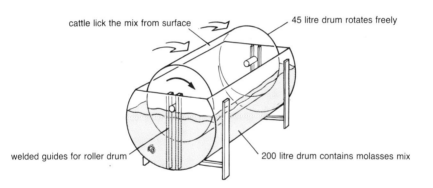

cattle lick the mix from surface

45 litre drum rotates freely

welded guides for roller drum

200 litre drum contains molasses mix

Fig 6.11   *Roller drum made from a 200 litre drum used for molasses and urea feed*
Fig 6.12   *Kenana cattle eating molasses, in Gezira, Sudan*

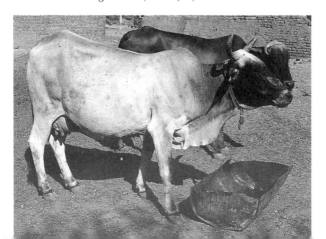

Mineral licks (Fig 6.13) could contain 60% salt (sodium chloride), 39% bone meal plus traces of sulphate, salts of cobalt, copper and iron, potassium iodide, magnesium and manganese. Cattle need calcium and phosphorus in a ratio of 2:1. Bone meal provides calcium and phosphorus in approximately this ratio, but if bone meal is not available dicalcium phosphate can be used. Calcium should be included in the lick at approximately 8% and phosphorus at 4%. When urea has been fed, an additional source of sulphur should be given to ensure that rumen microbes have an adequate supply. Sulphur is present in some proteins, but on low protein diets a deficiency may occur which could limit microbial action.

Nutritive content of feeds

Tables 6.1 and 6.2 show the nutrient content of some common tropical foodstuffs. Many more foods are available which are not included in the list, but those presented here indicate the range of values of foods available.

Fig 6.13 *Providing mineral blocks for Fulani pastoralists in Nigeria*

Table 6.1 Nutrient content of foodstuffs used for ruminants: cereal grains, supplements, concentrates and conserved forages

Foodstuff	DM* (g/kg)	ME* (MJ/kg DM)	CP* (g/kg DM)	Degradability (dg)
Cereal grains				
maize grain	860	14.2	98	0.65
sorghum grain	860	13.4	108	0.65
Supplements				
molasses	750	12.7	41	0.80
blood meal	900	13.2	942	0.50
meat and bone meal (high protein)	900	9.7	507	0.48
fish meal	900	11.1	701	0.29
Concentrates				
groundnut cake (decorticated)	900	12.9	504	0.74
soyabean meal (extracted)	900	12.3	503	0.83
Conserved forages				
maize silage	210	10.8	110	0.60
lucerne hay (before flowering)	850	8.3	193	0.78

* DM = dry matter
ME = metabolisable energy
CP = crude protein

Table 6.2 Nutrient content of foodstuffs used for ruminants: legumes and grasses

Foodstuff	DM* (g/kg)	ME* (MJ/kg DM)	CP* (g/kg DM)	Degradability (dg)
Pasture legumes				
Lucerne (*Medicago sativa*)				
early flower	240	8.2	171	0.80
inbud	220	9.4	205	0.80
before bud	150	10.2	253	0.80
Pasture grasses				
Rhodes grass (*Chloris gayana*)				
young leafy	270	9.3	190	0.80
early flower	300	8.7	80	0.80
full flower	350	8.9	74	0.80
Guinea grass (*Panicum maximum*)				
young leafy	250	8.0	208	0.80
flower	280	5.7	200	0.80
Napier grass (*Pennisetum purpureum*)				
young leafy	150	11.3	107	0.80
mid-flower	230	11.3	91	0.80
mature	270	9.3	52	0.80
Star-grass (*Cynodon aethiopicus*)				
young leafy	250	10.0	140	0.80
early flower	300	9.0	167	0.80
mature	400	8.3	83	0.80
Kazungula (*Setaria anceps*)				
young leafy	200	9.0	140	0.80
mid-flower	220	8.6	68	0.80
mature	270	7.8	52	0.80

Feeding stages in lactation cycle

Feeding the pre-partum cow

Feeding during the pre-partum period (**steaming-up**) is an essential part of feeding management. At calving the cow should have a body condition (see p. 68) of 3.0–3.5 and the farmer should aim to achieve this by feeding accordingly during the last six weeks of pregnancy, usually with high quality pasture or additional concentrate food up to a level of about 1% of body weight per day. Pastoralists and extensive graziers may have access to concentrate foods and strategically feed concentrates to those animals most in need – though this seldom if ever is the practice. The cow can be fed 2 kg/day concentrate at six weeks before calving and this amount should be increased steadily to calving, so that the cow receives 3–4 kg/day depending on body size.

In many systems where milk cows are fed natural pasture on extensively grazed land, cows calve at the beginning of the rains. Under these conditions it is difficult to achieve good pre-partum feeding.

In seasonal rainfall areas cows may lose weight in the dry season and show **compensatory weight gain** *when they are re-alimented on new pasture at the beginning of the wet season. Because of poor nutrition, such cows usually have low fertility, low milk yields and low lifetime milk production.*

Feeding the lactating cow

Early lactation

Peak milk yield occurs a few weeks after calving and the method of feeding must anticipate this. For low yielding cows, natural pasture in the early growth stage in humid regions may support lactation and keep cows in good condition. Later in the season some supplementation would usually be required if milk yield, body condition and fertility are not to suffer. It is unlikely that higher yielding cows will be able to eat enough roughage or pasture to sustain milk yield at this stage of lactation. Cows in early lactation can mobilise body reserves to meet some of the nutritional requirements of early lactation. This emphasises the importance of the cow's body condition at calving. imp .

Even for lower yielding cows, food must be given generously in early lactation to achieve maximum yield.

If possible in early lactation cows should be fed more than the calculated requirement. Milk yield can be determined by keeping *daily records*, but it must be assumed that the cow will produce more milk if fed more. Hence cows should be fed for what they might produce (**lead feeding**) and in more intensive systems using specialised dairy breeds, cows might need to be fed up to 2 kg food above the calculated requirement. Care should be taken, since this can produce fat cows without increasing milk yields. Experience helps guide the feeding regime at this stage of lactation, to determine which are the best ingredients to include in rations. In extensive systems, any additional food usually will be reflected in improved milk yield, fertility and body condition.

Mid-lactation

After the peak of lactation, yield declines at about 2.5% per week in dairy breeds. The yield of high producing cows may decline faster and the yield of heifers more slowly. Feeding should be adjusted according to milk yield and body condition should be monitored. The plane of nutrition is important at this stage if cows are to become pregnant again at the optimum time.

Late lactation

Lactating cows use the energy in food more efficiently than dry cows and this can be exploited in the latter part of lactation to replace weight lost earlier and to ensure that the cow is in good condition at calving. Whereas the cow may lose weight in early lactation (up to a total of 30 kg or 5% of body weight) she should gain 0.5 kg/day and up to 0.75 kg/day from mid-lactation onwards. If feeding is done properly, the need for high feeding levels in the dry period will be decreased. During the last three months of lactation, feeding should provide enough nutrients for both milk yield and some weight gain during mid-pregnancy. In the tropics, cows are often at their most vulnerable in the late stages of lactation, since this often coincides with the dry season when food resources are poor.

Nutrient requirements of cows

The topics discussed in the first part of Chapter 6 were general aspects of feeding milk producing animals, the foods available and some of the aims and problems that farmers face in trying to provide enough good quality food for their cows. The remainder of Chapter 6 describes the specific requirements of cattle for nutrients, particularly energy and protein, and explains how to calculate these requirements and formulate rations to meet the requirements. The equations and methodology used are derived from the metabolisable energy and protein systems described in the Agricultural Research Council (ARC), UK publications (1980 and 1984).

Energy partition in ruminants

Once food is eaten by the ruminant, it is broken down (or digested). Some of the energy in the food (**gross energy**) is lost in faeces, urine, methane and as heat. The gross energy is partitioned into **digestible energy**, **metabolisable energy** and **net energy** (Fig 6.14).

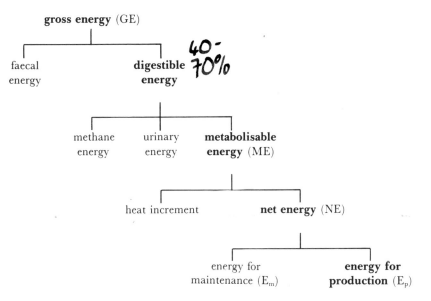

Fig 6.14 *Partition of food energy in ruminants*

The digestible energy can account for up to 70% of gross energy, depending on the food, but for poor quality roughages can be as low as 40%; the metabolisable energy is approximately 82% of digestible energy; the net energy is 60–90% of the metabolisable energy depending on the diet. The **heat increment** is heat which is lost as a result of the metabolic activities associated with the utilisation of food once absorbed and transported to the tissues. Some nutrients derived from the food require more digesting and metabolic processing than others and this produces more heat. Poor quality roughages have high heat increments, whereas concentrate foods have low heat increments associated with their utilisation in the body. The net energy can therefore vary between 20% and 50% of the gross energy of the food.

The unit of energy used is the **megajoule (MJ)** which is 1 million joules (J). The more familiar, now obsolete, unit of energy is the calorie (1 cal = 4.2 J). The energy requirements and allowances for ruminants are usually quoted as **megajoules of metabolisable energy per day (MJ ME/day)**. Energy requirements and allowances are calculated in terms of metabolisable energy.

Rumen fermentation

The rumen microbes (fungi, bacteria and protozoa) bring about the initial breakdown of the food eaten by the cow. The first microbes to attack the food are the fungi, followed by bacteria and protozoa. The microbes require a source of energy and nitrogen. They do not require a source of dietary protein and a ruminant could, if necessary, survive without protein in the diet, if a source of available nitrogen was fed. Under normal circumstances this would not be advisable or necessary. The microbes themselves are a source of protein for the ruminant animal and are broken down and absorbed in the small intestine. The complex and simple carbohydrates in the cow's diet are broken down by microbes. Volatile fatty acids are produced as a result and are absorbed across the rumen wall. These acids are the main source of energy for the ruminant animal. The main volatile fatty acids are acetic, butyric and propionic acids. Acetic acid has its highest concentration on roughage diets, while propionic acid has its highest concentration on concentrate diets. Propionic acid is the main precursor of glucose for the ruminant and hence is important for milk producing animals.

imp. for milk (handwritten margin note)

> *The rate of rumen fermentation depends mainly on the quality of food ingested. The rate of food breakdown determines the rate of passage of food through the digestive tract, which in turn influences the rate of food intake.*

Dry matter intake

If a food is palatable to a cow, then the amount of a good quality, non-bulky (low roughage) food eaten – the dry matter intake (DMI), depends largely on the quality of the food and the physiological state of the animal. An animal would usually eat a greater quantity of a non-bulky food compared with a bulky food. For bulky foods, intake depends on how quickly the food passes through the alimentary tract (i.e. the rate of passage of the food), which depends on digestibility. For optimum food intake, an animal must also have an adequate supply of water, since dehydration can reduce food intake. High ambient temperatures also reduce food intake by causing a reduction in the secretion rate of the hormone thyroxine which plays a part in controlling appetite.

The quality of food is described as the **energy concentration** or **energy density** – the amount of energy as a unit weight of the dry matter of the food. The units of energy concentration are megajoules of metabolisable energy per kilogram of dry matter (**MJ ME/kg DM**). The energy concentration is also known as the M/D (M over D) – the metabolisable energy in the dry matter (MJ ME/kg DM). Quality is also described by the term **metabolisability (q)** derived from metabolisable energy divided by gross energy (q = ME/GE MJ/kg DM).

A cow will usually eat dry matter (DM) equivalent to between 2.5% and 3% of her body weight each day. For a 400 kg animal this would be 10–12 kg DM/day or 3650–4380 kg DM/year. For a pasture grass with 30% DM, this would amount to up to 14.5 tonnes/year fresh herbage. Higher dry matter intake will occur at higher nutrient densities. Intake is also related to level of production and pregnancy status. High yielding cows will eat more than low yielding cows; pregnant cows will eat more than non-pregnant cows. If necessary, the intake of lactating cows can be estimated using an equation, which relates intake to body weight, level of production and diet quality: DMI = 0.025 BWt + 0.1Y (kg DM/day) where BWt = body weight (kg) and Y = milk yield (kg). The probable dry matter intakes of cows can be calculated from the data shown in Table 6.3.

Table 6.3   Probable dry matter intakes of cows in mid- and late-lactation

Body weight (kg)	Dry matter intake (kg DM/day)				
	Milk yield (kg)				
	5	10	15	20	25
350	9.3	9.8	10.3	10.8	11.3
450	11.8	12.3	12.8	13.3	13.8
550	14.3	14.8	15.3	15.8	16.3

Energy requirements

The metabolisable energy (ME) requirements for dairy cows can be calculated easily, but there is no need for them to be calculated every time. Tables 6.4, 6.5, 6.6 and 6.7 and the following equations summarise the ME allowances for maintenace, growth, lactation and pregnancy calculated according to the methods presented by ARC (1980 and 1984).

The maintenance metabolisable energy allowance (M_m) is most easily calculated using the equation $M_m = 8.3 + 0.091 \times BWt$. Using this equation the maintenance allowances (MJ/day) (Table 6.4) can be derived for cows of different weights. These allowances:

1 Have a built-in safety margin of 5% to take into consideration any variation in food quality.
2 Have a 10% activity allowance to account for normal daily activity (but not draught power production).
3 Assume that ME is used with an efficiency for maintenance (K_m) of 0.72 (dietary energy is used with different efficiencies for different uses within the body).

Table 6.4 Metabolisable energy allowances for maintenance (ME_m)

Body weight (kg)	ME_m (MJ/day)
100	17
150	22
200	27
250	31
300	36
350	40
400	45
450	49
500	54
550	59

The efficiency of utilisation of ME for liveweight gain and fattening (K_f) varies considerably with different types of food. These variations can be related to the energy concentration of the ration and K_f may be calculated using the equation: $K_f = 0.0435$ M/D. K_f varies (0.30–0.60), as M/D varies (7–14 MJ/kg DM). The extra allowance for growth can be smaller on better diets, since the cow will utilise energy more efficiently. The additional allowances for growth (Table 6.5) take this into consideration.

In lactating cows body tissue is laid down with a higher efficiency than in non-lactating animals and this is not related in this case to quality of the diet. Dietary ME of 34 MJ is required for every 1 kg of liveweight gain (LWG) made by a lactating cow and this should be added to the ME allowance.

If a cow is losing weight during lactation, the energy produced from the tissue mobilisation is set against the ME requirement. Body tissue has an energy value of 20 MJ/kg and is used with an efficiency of 0.82 to produce milk containing 16.4 MJ of energy. This is equivalent to a dietary ME intake of 16.4 × 1.05/0.62 MJ (including a 5% safety margin and assuming that ME is used with an efficiency of 0.62 for milk synthesis). Thus is a cow is losing 1 kg bodyweight each day, then the allowance can be reduced by 28 MJ/day.

Table 6.5 Metabolisable energy allowances for growth in non-lactating cows

Body weight (kg)	M/D (MJ/kg DM)	ME growth allowance (MJ/day)			
		Growth rate (kg/day)			
		0.25	0.50	0.75	1.00
300	8	10	21	–	–
	10	8	17	28	–
	12	7	14	23	34
400	8	11	25	–	–
	10	9	20	32	–
	12	8	16	29	40
500	8	12	28	–	–
	10	10	22	37	–
	12	8	19	31	45

Table 6.6 Metabolisable energy allowance for pregnancy

Month of pregnancy	ME pregnancy allowance (MJ/day)
<6	<5
6	8
7	11
8	15
9	20

Pregnancy requires small amounts of energy during the first 6 months, but during the last 3 months appreciable amounts of energy are required to support the growth of the fetus and conceptus (Table 6.6).

The metabolisable energy allowance for milk production (M_1) depends on the efficiency with which ME is used for lactation ($K_1 = 0.62$), the energy value of the milk EV_1) in MJ/kg and the milk yield. If a 5% safety margin is included, then $M_1 = 1.05 \times EV_1/0.62$ MJ/kg milk. Using this equation and the equation $EV_1 = 0.0386$ BF $+ 0.0205$ SNF $- 0.236$ (where BF = butterfat, SNF = solids-not-fat, units g/kg), then the values for ME allowances for milk production shown in Table 6.7 can be derived.

The efficiency of use of metabolisable energy for different production functions varies and increases with increasing quality of the diet. Energy is used most efficiently for maintenance, followed by lactation and growth (Table 6.8).

Table 6.7 Metabolisable energy allowances for lactation (M_l)

SNF (%)	M_l (MJ ME/kg milk per day)						
	BF (%)						
	3.4	3.8	4.2	4.6	5.0	5.4	5.8
8.4	4.74	5.00	5.26	5.52	5.79	6.05	6.31
8.8	4.88	5.14	5.40	5.66	5.92	6.19	6.45
9.2	5.02	5.28	5.54	5.80	6.06	6.73	6.59

Table 6.8 Efficiency of use of metabolisable energy

Body functions	M/D			
	7.4	9.2	11.0	12.9
Maintenance (K_m)	0.64	0.68	0.71	0.75
Growth (K_f)	0.32	0.40	0.48	0.55
Lactation (K_l)	0.56	0.60	0.63	0.67

Source: Macdonald, Edwards and Greenhalgh (1987)

Protein requirements

Protein requirements are calculated according to the demands of nitrogen made by the rumen micro-organisms and for protein by the ruminant animal itself. The system described here is known as the **metabolisable protein system** and has replaced the digestible crude protein system. The proposals for calculating protein requirements (i.e. the metabolisable protein system) given by ARC (1980 and 1984) however, are under review and the quantitative aspects are not finalised.

The protein found in ruminant foods can be divided into two components:

1 **rumen degradable protein (RDP)** broken down (degraded) in the rumen and used by micro-organisms
2 **undegradable protein (UDP)** cannot be broken down by microbes and passes from the rumen to be digested in the small intestine.

The calculation of the requirements for RDP and UDP starts with the ME requirement for energy and assumes that the energy requirements have been met by the diet. The rumen degradable protein requirement is then calculated by applying the formula: RDP requirement = 8.4 × ME requirement (g/day).

This also equals the amount of protein provided by the microbes – **microbial protein (MP)**, to the ruminant animal. The value of the protein in microbial protein depends on its actual amino acid content

(80%), the amount digested in the small intestine (80%) and the amount of amino acids actually absorbed from the small intestine (85%). These values are approximate averages. The microbial protein supplied would therefore equal:

1   $MP = 8.4 \times ME \times 0.8 \times 0.8 \times 0.85$
2   $MP = 4.57 \times ME$ (g/d)

The better the diet the more microbial protein will be produced. It is necessary also to consider the **essential amino acids (EAA)** in the protein. The proportion of essential amino acids in the microbial protein in the digestive tract is 0.48. Therefore, essential amino acids supplied by the microbes is: $EAA = 4.57 \times ME \times 0.48$ g/day.

The protein needed by the ruminant animal is calculated by adding-up the requirements for maintenance, growth, lactation and pregnancy. Additional amounts may be required for activity and hair growth. The total is known as the **tissue protein (TP) requirement**. Again it is better to consider essential amino acids. The EAA content of milk is 0.53 and therefore, EAA required = TP × 0.53.

If the microbial protein (MP) is not enough to meet the tissue protein requirement (TP) then undegradable protein (UDP) is required in the diet, the amount calculated by $UDP = (TP - MP)/(0.8 \times 0.85)$, again assuming 80% digestion and 85% absorption. The degradability (i.e. the proportion which is degraded) required in the dietary protein is calculated by the equation: degradability (%) = [RDP/(RDP + UDP)] × 100.

Water requirements

Water intake depends on food intake, nature of the diet, physiological state of the animal and ambient temperature (Table 6.9). Indigenous

Table 6.9 Optimum water intake depending on ambient temperature for dairy cows in warm climates

Milk yield (kg)	Ambient temperature			
	11–20°C		>20°C	
	Body weight (kg)			
	350	600	350	600
0	46	63	56	77
10	58	86	70	105
20	69	98	84	119
30	81	109	98	133
40	98	120	119	147

Water intake = liquid + food moisture kg/day
Source: Oliver (1987)

breeds may drink less than exotic breeds, because they are better adapted to hot environments and they have a lower body water turnover. Ideally a constant supply of water should be available, but if stock have to be taken each day to water (i.e. the river or a dam) exotic breeds and crosses would suffer more than local stock. Extensively managed, indigenous stock are often watered once a day, but this would not be recommended for milking cows.

Mineral requirements

The quantity of minerals required for maintenance depends on body size. The most important minerals as far as quantity is concerned are calcium, phosphorus and magnesium. Additional amounts of minerals are needed for growing, for pregnant and lactating cows.

Legumes and grains contain large amounts of calcium and phosphorus respectively, compared with other foods. The adult as well as the growing animal requires a constant intake of calcium. Milk fever is caused by inadequate calcium levels in higher yielding cows. A 500 kg animal producing 18 kg milk would require about three times as much calcium and phosphorus as a non-lactating, non-growing animal. The calcium and phosphorus requirements for 18 kg milk/day are similar to those for growth of 1 kg/day (Table 6.10).

Phosphorus in herbage is related to protein content and is likely to be low in the dry season. The optimum ratio in the diet of calcium and phosphorus is between 1:1 and 2:1. The metabolism of calcium is related to that of phosphorus and even if neither element is deficient, an excess of either will affect the metabolism of the other.

Table 6.10 Calcium, phosphorus and magnesium requirements for maintenance and growth

Body weight (kg)	Growth rate (kg/day)								
	0.0			0.5			1.0		
	Mineral requirements (g/day)								
	Ca	P	Mg	Ca	P	Mg	Ca	P	Mg
100	2	2	2	12	6	3	21	11	4
200	5	3	4	14	8	5	24	13	6
300	7	5	5	17	9	7	26	14	8
400	9	8	7	19	15	8	28	21	9
500	12	10	9	21	17	10	30	23	11
600	15	12	11	23	19	12	32	25	13

Ca = calcium, P = phosphorus and Mg = magnesium

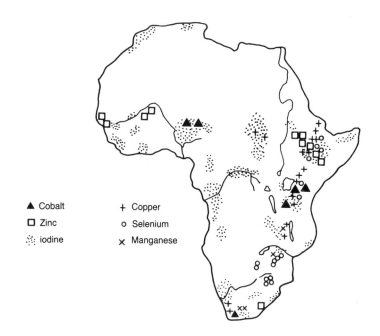

Fig 6.15 *Geographical distribution of mineral deficiencies, in Africa*

Other minerals required by milking cattle include sodium, chlorine, potassium and sulphur and trace elements such as iron, iodine, cobalt, copper, molybdenum, selenium, fluorine, zinc and manganese. Deficiencies may occur depending on the feeds and the soil conditions where cows depend largely on grazing. A deficiency of salt (sodium chloride) can cause poor growth and reduced appetite. Salt licks should be provided to ensure food intake.

Mineral levels in soil, grass and animal tissues, as well as the year or season, are variables which must be taken into account to assess the mineral status of an area. A summary of the geographical distribution of certain mineral deficiencies in Africa is shown in Fig 6.15.

Vitamin requirements

Vitamins are

1 fat soluble – Vitamins A, D, E and K
2 water soluble – Vitamins B complex and Vitamin C.

Under normal grazing conditions tropical cattle should not suffer from vitamin deficiencies. In the dry season there is concern that Vitamin A might be deficient, since its precursors carotene and carotenoids are low in dry vegetation, but enough Vitamin A is usually stored in the liver to

109

maintain the cow through the dry season. Young calves may suffer from Vitamin A deficiency if born late in the dry season, since the colostrum is normally Vitamin A deficient at this time. It takes about a month for levels to be replaced in the liver. Vitamin D is synthesised in the skin by the action of sunlight and animals should not become deficient unless housed. Vitamin E (tocopherol) is widely distributed in animal feeds. The animal can store Vitamin E well and good pre-partum feeding leads to a high level in colostrum and milk. Vitamin K, which is concerned with the blood clotting process, is synthesised by rumen microbes. The B complex vitamins are also synthesised by rumen microbes and are unlikely to be deficient except in the first two months of life of the calf or in very high yielding dairy cows. Vitamin B_{12} requires cobalt for its synthesis and hence in areas where soil cobalt deficiency occurs, this vitamin also may be deficient.

Table 6.11   Ration requirements of 400 kg cow producing 10 kg milk/day

Requirement	Amount (per day)
ME	
for maintenance ME	45 MJ
for milk ME	63 MJ
TOTAL ME requirement	108 MJ
Protein	
for RDP	907 g
but Microbial Protein (EAA) produced	237 g EAA
for milk (30 × 10)	300 g
for maintenance (2188 mg/kg$^{0.75}$)**	196 g
for hair (112.5 mg/kg$^{0.75}$)	10 g
TOTAL EAA requirement	268 g EAA
but EAA deficit not supplied by rumen microbes	31 g EAA
for UDP	46 g
TOTAL CP requirement	953 g CP
protein degradability required	95%
minimum CP content required in the diet	87 g CP/kg DM

* The cow is assumed to be 400 kg liveweight, such as a White Fulani or Sahiwal, producing 10 kg milk/day of 5.8% BF and 8.4% SNF, in early lactation and at a constant weight.

** kg$^{0.75}$ = metabolic weight (see *Glossary*)

Note   ME = metabolisable energy, RDP = rumen digestible protein, EAA essential amino acids, UDP = undegradable protein and CP = crude protein. Energy and protein requirements based on methods presented earlier (pp. 100–107).

Ration formulation and feeding in practice

The methodology described to determine the nutrient requirements of ruminants is complex, but a knowledge of the factors involved is important in order to understand the ruminant's digestive functions and how the animal's food is used to meet its nutrient requirements. Making rations for dairy animals uses this methodology to determine the animal's requirements. Once this is known, the farmers must consider the foods available and mix these in a way that meets the requirements as nearly as possible. This may require some 'trial and error'.

The dry matter intake requirements are calculated using the equation DMI = 0.025 BWt + 0.1Y which does not take into account the quality of the diet. The DMI required of 11 kg DM/day is considered to be a maximum. DMI may also be predicted using the equation DMI = 106q + 37P + 24.1 (where q = metabolisability and P = proportion of concentrate in the diet), which predicts a lower intake).

Ration formulation

Based on these requirements (given above) and knowing the nutrient content of available feeds, it is possible to formulate rations. It is possible that the feeds available may not be able to support the level of milk production required (i.e. 10 kg) – in this case a ration is formulated and an estimate made of the milk yield that it will support. These calculations assume that the feeds do not contain any toxic factors which would inhibit intake.

The DM, ME, RDP and UDP of the individual ingredients of the proposed ration are determined from feed tables (Table 6.12). The feeds used in this example are natural pasture grasses, Napier grass, Stylo, Rhodes grass (young leafy and early flower stages), leucaena, lucerne hay, maize silage, groundnut cake, cottonseed cake, maize grain, sorghum grain, molasses and urea.

Diets

Nine diet formulations are presented here using different combinations of these feeds. The nutrient contents of feeds are average; in practice the nutrient contents will vary from average figures depending on locality and season. The formulations are intended to indicate the **principles of formulation** and should not be taken as recommended diets using these ingredients. In practice a diet should be formulated and then fed and adjusted according to animal performance.

Table 6.12 Nutrient content of feed ingredients

Feed ingredients	DM (g/kg)	ME MJ/kg DM	q*	CP g/kg DM	dg**	RDP g/kg DM)	UDP (g/kg DM)
Natural pasture							
wet season	300	8.0	0.43	90	0.80	72	18
dry season	550	5.5	0.30	28	0.80	22	6
Napier grass							
early flower	230	10.0	0.54	90	0.80	72	18
late flower	270	9.3	0.54	52	0.80	42	10
Rhodes grass							
young leafy	270	9.3	0.54	190	0.80	152	38
early flower	300	8.7	0.47	80	0.80	64	16
Stylo	300	9.0	0.49	150	0.80	120	30
Leucaena	300	10.7	0.58	230	0.65	160	70
Lucerne hay	850	8.3	0.45	193	0.78	151	43
Maize silage	210	10.8	0.58	110	0.60	66	44
Groundnut cake	900	12.9	0.70	504	0.74	373	131
Cottonseed cake	925	10.9	0.59	360	0.50	180	180
Maize grain	860	14.2	0.77	98	0.65	64	34
Sorghum	900	12.0	0.65	100	0.65	65	35
Molasses	750	12.7	0.69	42	0.50	21	21

* q = metabolisability

** dg = degradability of crude protein

The total DM, ME and RDP/UDP for each diet are compared with the calculated requirement. The ME contents of the diets are calculated to be as close as possible to the cow's requirement for the predicted DM intake. If there is an RDP deficit this can be corrected by increasing one of the ingredients or by increasing the non-protein nitrogen content of the ration using urea. Urea nitrogen is assumed to be converted to microbial nitrogen with an efficiency of 0.80. The amount of urea (g/day) required = deficit in $RDP/(6.25 \times 0.80 \times 0.46)$ (see p. 106). If there is a deficit in both RDP and UDP, the RDP deficit should be corrected first. If there is an excess of RDP or UDP, a new ration can be formulated to reduce the excess, but this assumes that the energy requirements have been met by the diet.

> *For all the rations described, the dry matter values for each diet component must finally be converted to fresh weights to determine the amount of the fresh food which must be fed.*

Diets based on natural pasture
Natural pasture is good quality at the beginning of the rainy season, but quality soon falls off as the grass gets older. Diet 1 is based entirely on natural pasture/roadside grasses and Diets 2 and 3 are based on supplemented pasture:

Diet 1

Feed	DMI (kg/day)	ME (MJ/day)	CP (g/day)	RDP (g/day)	UDP (g/day)
Predicted requirement	11.0	108	953	907	46
Natural pasture wet season	8.4	67	756	605	151

Diet 1 indicates that the natural pasture in this example would not support a 400 kg cow producing 10 kg milk/day. The energy content of the pasture might support 4 kg milk.

Cows grazed on roadside grass and cut-and-carry forage could be supplemented with grown forage, such as Napier grass and leucaena as shown in Diet 2. The energy of this diet would support a 400 kg cow producing approximately 6 kg milk/day. Excess protein would be used as a source of energy.

Diet 2

Feed	DMI (kg/day)	ME (MJ/day)	CP (g/day)	RDP (g/day)	UDP (g/day)
Predicted requirement	11.0	108.0	953	907	46
Roadside grass					
wet season	4.3	34.0	387	310	78
Napier grass	2.0	18.6	104	84	20
Leucaena	3.0	32.1	690	480	240
TOTAL diet	9.3	84.7	1181	874	307

Alternatively, natural pasture in the dry season could be supplemented with a fodder legume, such as stylo from a fodder bank and with molasses and urea as shown in Diet 3. This diet would support a 400 kg cow producing 5 kg milk/day.

Diet 3

Feed	DMI (kg/day)	ME (MJ/day)	CP (g/day)	RDP (g/day)	UDP (g/day)
Predicted requirement	11.0	108.0	953	907	46
Natural pasture					
dry season	2.9	16.0	81	64	17
Stylo	4.0	36.0	600	480	120
Molasses	2.0	25.4	84	42	42
Urea	0.1	0	266	266	0
TOTAL diet	9.0	77.4	1029	852	179

Diets based on conserved forages

Diet 4, based on fresh maize silage (10 kg DM/kg of silage at 210 g DM/kg fresh silage = 48.6 kg/day of fresh silage), would provide enough energy for 10 kg milk, but not enough RDP and too much UDP. Even so, the CP concentration in the DM is greater than required.

Diet 4

Feed	DMI (kg/day)	ME (MJ/day)	CP (g/day)	RDP (g/day)	UDP (g/day)
Predicted requirement	11	108	953	907	46
Maize silage	10	108	1100	660	440

By substituting some of the silage for lucerne hay, the energy and RDP requirements can be met, but the CP concentration is still too great and too much UDP is provided, as shown in Diet 5. Excess protein would be broken down and used as an additional source of energy.

Diet 5

Feed	DMI (kg/day)	ME (MJ/day)	CP (g/day)	RDP (g/day)	UDP (g/day)
Predicted requirement	11.0	108	953	907	46
Maize silage	8.0	86	880	528	352
Lucerne hay	2.6	22	502	393	112
TOTAL diet	10.6	108	1382	921	464

An alternative to lucerne hay would be to supplement maize silage with urea. The RDP deficit in Diet 4 is 247 g. The amount of urea needed is 247/2.3 = 107.4 g/day. This would be a safe amount to feed with 47.6 kg maize silage. The decision regarding which diet to chose would depend on relative costs and availability of the feeds (i.e. the costs of urea and lucerne hay and the quantity of silage available). In Diet 5 only 38 kg fresh silage was used compared with 48 kg for Diet 4. If lucerne hay is cheaper than urea and available then this might be preferred to urea.

Diets based on sown pasture

If cows were grazed on young leafy and early flowering Rhodes grass pastures, this would provide nutrients as shown in Diets 6 and 7.

Diets 6 and 7

Feed	DMI (kg/day)	ME (MJ/day)	CP (g/day)	RDP (g/day)	UDP (g/day)
Predicted requirement	11.0	108	953	907	46
Rhodes grass					
Diet 6: young leafy	9.9	92	1880	1504	376
Diet 7: early flowering	8.9	77	712	570	142

Ad lib. intake of young leafy pasture meets the animals' energy requirements for maintenance and 7 kg milk, but at the later growth stage, the pasture would only meet the requirements for 5 kg milk. The protein

115

content at the later stage is greatly reduced and might also limit milk production.

From Diets 1, 2, 3, 6 and 7 it can be concluded that pasture will not provide the full nutrient requirements for a 400 kg cow to produce 10 kg milk and that some additional foods will have to be provided. When pasture has a lower nutrient value (i.e. more mature or in the dry season) then this will support lower production and will require greater supplementation. It can generally be stated that sustained milk production for a full 305-day lactation at levels above 3 kg/day will require food supplements in addition to pasture. Dairy development programmes which do not provide methods for the provision of food inputs in addition to pasture are unlikely to succeed. If farmers cannot, or are not willing to provide additional food for dairy cows, they will not be able to increase milk yields above the minimum levels normally achieved by these cows.

In Diet 6, even if the animal could eat enough young Rhodes grass, this would still undersupply the energy requirements, but some RDP and UDP would be surplus. This protein would be used as an additional energy source and would help to support lactation. Some supplementation would be necessary later in the season when the pasture is in the early flower stage, as shown in Diets 8 and 9.

Diet 8 Rhodes grass

Feed	DMI (kg/day)	ME (MJ/day)	CP (g/day)	RDP (g/day)	UDP (g/day)
Predicted requirement	11.0	108	953	907	46
Rhodes grass early flower	7.7	67	616	493	123
Maize grain	2.0	28	196	128	68
TOTAL diet	9.7	95	812	621	191

Diet 9 Rhodes grass

Feed	DMI (kg/day)	ME (MJ/day)	CP (g/day)	RDP (g/day)	UDP (g/day)
Predicted requirement	11.0	108	953	907	46
Rhodes grass early flower	7.7	67	616	493	123
Groundnut cake	2.0	26	1008	746	262
TOTAL diet	9.7	93	1624	1239	385

116

In the early flowering stage Rhodes grass cannot provide the requirements of the animal (Diet 7). If a maize grain supplement is given, this improves the energy supply, but the RDP supply is still low (Diet 8). By supplementing with 2 kg groundnut cake, a similar energy supply is provided and excess RDP and UDP would be used to provide energy. Again, the best option might be to supplement the maize grain with molasses/urea. The RDP deficit for 10 kg milk in the Diet 8 with maize grain is 286 g/day. The urea required is therefore $286/2.3 = 124$ g/day. With 1 kg molasses, this would bring the diet to the requirement for 10 kg milk.

7 Milk breeds and breeding for milk

Descriptions of the cattle breeds in the world have been presented by Felius (1985) and Mason (1988). More recently, tropical cattle have been described by Maule (1990) who listed approximately 200 well defined breeds and numerous non-descript cattle. Some preferred breeds for milk and some specialised dairy breeds are discussed in the following sections.

> *There are few distinct dairy breeds in the tropics, but some breeds are preferred for milk production. Many breeds are dual-purpose or multi-purpose and fulfil a variety of roles as producers of milk, draught power, meat, dung and hides. No breed is suited to all circumstances and where there is a choice, farmers must find the breed most suitable to the environmental conditions and husbandry which they can offer their animals.*

Cattle are classified or morphological grounds into two species – *Bos indicus* (humped or Zebu cattle) and *Bos taurus* (humpless cattle). Individuals from these two groups can interbreed to form fertile offspring and are not considered to be different species in biological terms. In East and southern Africa, Sanga cattle are found which result from crosses between Zebu and humpless longhorn cattle of Egypt. *Bos taurus* found in the tropics are mainly those imported from Europe, such as the Friesian, Ayrshire and Jersey, though the N'Dama breed (West Africa) is a taurine type. *Bos taurus* cattle also include Criollo cattle in Latin America.

Breeds of milk cattle
Indigenous breeds

Approximately 30 breeds are found in the Indian sub-continent, of which the **Sahiwal**, **Red Sindi** and **Gir** are the better milk producers.

118

Fig 7.1 *Sahiwal and Sahiwal × Friesian cattle, at Naivasha, Kenya*

Fig 7.2 *Sokoto Gudali bull, in Nigeria*

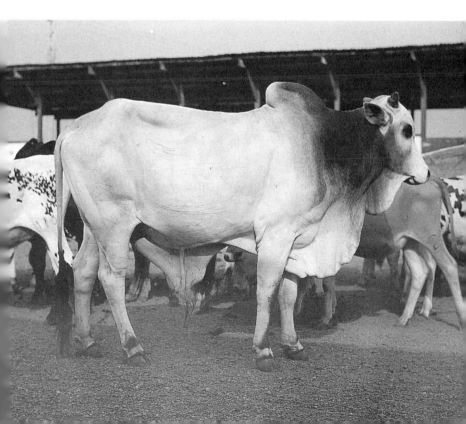

Fig 7.3 *Boran cattle, in Kenya*

Fig 7.4 *Butana cattle, in Sudan*

West African breeds include the **White Fulani** (**Bunaji**) which are large well muscled animals with thoracic humps, lyre-shaped horns and are white with black points and the **Sokoto Gudali** which is kept by the Fulani and Hausa people for draught power and milk production.

East African and southern African Zebu breeds include the **Boran**, **Butana**, **Kenana**, small **East African Zebu** breeds, such as the **Nandi**, **Maasai**, **N'Kedi** and **Lugware** and the **Malawi Zebu**.

African Sanga breeds include the **Ankole** from Uganda, Zaire, Rwanda and Burundi and the **Barotse** kept by the Barotse and Lozi people of south east Angola and west Zambia in the Zambezi flood plain area.

Latin American and Caribbean breeds include the **Costa Rican Dairy Criollo**, **Rio Limon Dairy Criollo**, **Blanco Orejinegro** (BON) (Black-eared White) and **Jamaican Hope**.

Specialised dairy breeds

European breeds (**Ayrshire**, **Shorthorn**, **Guernsey** and **Jersey** bulls) were introduced in the eighteenth century to the India sub-continent to several military farms. Since then, these and other high yielding breeds including **Brown Swiss**, **Friesian** and **Danish Red** have been imported and crossed with hardy, adapted native Zebu cattle to produce good dairy breeds. Under systems of intensive feeding and management, the

ig 7.5 *Kenana cattle, at Kosti, Sudan*

121

Fig 7.6   *East African Zebu cattle being used for ploughing, in Kenya*

Friesian-Holstein cross, irrespective of the indigenous breed and agro-ecological situation has proved to be the best exotic cross followed by Brown Swiss and Jersey.

Although productive indigenous breeds of milk cattle exist throughout the tropics, the main strategies for increasing milk production in tropical cattle have been dominated by large superiority in **additive genetic merit** for milk yield of temperate breeds over tropical breeds. First generation crossbreds generally have yields twice that of the indigenous purebreds. As a result, European breeds have been introduced to many parts of the tropics and crossbreeding has produced a number of stabilised crossbreds.

At the same time there is an increasing interest in the preservation of indigenous breeds and their improvement by selection, but such work is likely to proceed in parallel with improvements by the introduction of temperate sires. For long term breeding plans, the creation of a new breed for crossbred foundations is the best option.

A number of examples of **new breeds** exist. These are not numerically important and none have gained a reputation as a major dairy breed. The **Karan Swiss**, **Karan Fries**, **Sunandini**, **Mpwapwa** and the **Australian Milking Zebu** are some examples of new breeds.

Purebred exotic *Bos taurus* cattle, such as the Holstein-Friesian, Ayrshire and Jersey, have been introduced into many countries.

122

Fig 7.7 *Australian Milking Zebu cows, in Malaysia*

Breeding and genetic improvement

For a more comprehensive insight into breeding and improvement, see *Animal Breeding* by Gerald Wiener in *The Tropical Agriculturalist* series. If management, husbandry and feeding are good and environmental constraints overcome, the milk yield will approach the maximum yield the cow is capable of. Both within herds and at the national level, in most tropical countries there is a great potential for increasing milk yield by a combination of improved husbandry and improved breeding.

Improved genetic merit can be achieved in one of three ways:

1 selection within the local stock for increased production
2 by crossing local stock with other breeds of indigenous or exotic cattle
3 by replacing local cows with other breeds.

The second method is usually adopted, since buying improved cows is expensive and such cows are not usually sold by farmers who have them. Farmers commencing a new milk production enterprise will wish to begin with the most suitable breed, but will usually begin with local cattle and then **upgrade** these by one of the above methods.

Culling and selection

In larger herds when the level of management and husbandry are good, some improvement of milk yield might be possible by increasing the genetic merit of the stock through culling and selection. Slow genetic improvement can be brought about by removing poor milkers from the herd and replacing them with better milk producing cows. The degree to which improvement by this method can be achieved depends on the **heritability** of the traits for which selection is made. The heritability of milk producing traits is relatively low. Heritability (h^2) is defined as:

123

1 the proportion of the phenotypic variation (i.e. that which can be seen and measured in the animal population) attributable to additive genetic effects (i.e. variation is controlled by a number of genes rather than by a single gene);

2 a measure of the extent that a cow or bull can pass on its characteristics to its offspring.

Estimates of heritability vary from 0.0 to 1.0. When $h^2 = 0.0$, the trait is not heritable under the environmental conditions but at $h^2 = 1.0$, the trait would be fully heritable under the prevailing conditions.

Greatest progress can be made when the number of cows being selected is large. A farmer with only one or two cows has little opportunity to influence the genetic merit of future cows by culling or selection. Such a farmer may have one female calf born once a year and will have to keep all female calves born either to replace the older cows or to allow the herd to grow. If a choice arises to replace a poor producer cow with the daughter calf of a better producer, this option should be taken. If cows are poor producers only because they have suffered from a disease such as mastitis, then their daughters might also be worth keeping. Records and the farmer's experience will help to make such a decision. Without good records, improvement is difficult to achieve. Selecting from a greater number of animals allows the **selection intensity** to be greater. **Selection indices** based on production traits and aspects of economic value can be used to aid selection.

If possible, poor milk producers should be culled and their daughters should not be used for future milk production. The larger the herd, the greater the potential for overall improvement by culling and replacing with better stock. Genetic improvement by selection is slow even when numbers are high.

Having good records can assist the farmer to identify good milk producers and to improve the quality and milk production of stock.

Crossbreeding and grading-up

The quickest way to achieve improvement of milk yield by genetic improvement is by crossbreeding with a breed that has a higher genetic potential to produce milk. Such a breed is often called an improved breed or, because it comes from somewhere else, an exotic breed. These terms are misleading, because breeds which do well in some environments may do less well in others. Hence exotic breeds do not necessarily perform better than local breeds and are not necessarily an 'improvement'.

Increased production by crossbreeding is most easily achieved through the male line by having an 'improved bull' or by using AI. With the advance of the technique of multiple ovulation and embryo transfer (MOET), rapid improvement through the female line is now feasible.

An important benefit from crossbreeding is **hybrid vigour** – the half-bred is more vigorous and able to survive better than the parents.

If crossbreeding is to be undertaken, the farmer must be able to ensure that a high level of management and husbandry will be provided for the crossbred stock. Once management and husbandry are sufficiently good, crossbreeding can be undertaken to produce cows with a greater proportion of improved blood. A bull, such as a Sahiwal, Jersey or Friesian, may be crossed with a local cow. The first cross will be a 50/50 grade animal. Subsequent crosses to one of the parent breeds will be genetically closer to that breed.

A problem arises in crossbreeding, since the second cross will be a back-cross to one of the parent breeds to produce a 75/25 cross. Maintaining 50/50 crossbred populations is difficult and the farmer must choose a method of crossbreeding to achieve the best grade animal for the environment. If a 75% improved breed is too much, then the local breed should be used to give a 75/25 local/improved cross. This can then be crossed with the improved breed to give a 62.5/37.5 mix. Crossbred bulls could be used.

Criss-cross breeding consists of using an exotic bull in one generation and a local in the next (Fig 7.8). In **three way crossing**, alternation between three (or more) breeds of bull occurs.

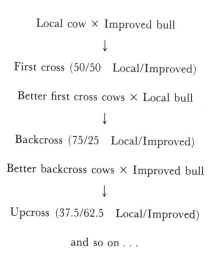

Local cow × Improved bull

↓

First cross (50/50   Local/Improved)

Better first cross cows × Local bull

↓

Backcross (75/25   Local/Improved)

Better backcross cows × Improved bull

↓

Upcross (37.5/62.5   Local/Improved)

and so on . . .

Fig 7.8   *An example of criss-cross breeding*

Artificial insemination (AI)

The fastest way to bring about genetic improvement to a herd is by using AI with bulls of high genetic merit for milk production. AI has the advantage over natural service that the transmission of venereal diseases from the bull to cow or heifer is avoided if care is taken to use healthy semen donors. If semen is to be imported from bulls tested in temperate environments, it is important to import from a number of bulls and to re-test in the tropical environment.

The bulls used for AI need to be of known superior merit. They should be tested according to the merit of their daughters (progeny testing), but this can take up to eight years.

Progeny testing

Young dairy animals (bulls or heifers) can only be assessed by the performance of their parents or their older siblings (or half siblings). An older bull can be assessed by the performance of its daughters. Dairy bulls and bulls used for AI are often tested by this method. It takes six or seven years from the bull's birth before lactation results of his daughters are available. It must be remembered that most exotic bulls are tested in an environment very different from the tropics and may not perform as well in the tropics as in their original environment.

8 Milk products

When planning the development of a dairy industry, attention also must be given to methods of collecting milk from farmers and delivering fresh milk or milk products to consumers. The target consumers are often found in larger urban centres situated at a distance from producers. Milk hygiene, collection, transport, preservation, processing and distribution are therefore integral aspects of the dairy industry and dairy development programmes.

Milk hygiene

Good hygiene and sanitation practices must be carried out at all stages of milking, transport and processing. Attention should be given to the cleanliness of the animals, the milkers, the milking utensils and the transport containers, to the health of animals (control of the diseases tuberculosis, brucellosis and mastitis) and to the health of the milkers. At processing further contamination must be prevented, proper heat treatment carried out and cool storage provided if possible.

Methods of preserving milk

There is no substitute for good hygiene and this should be the first step towards keeping milk fresh. Clean, microbe-free milk will keep longer than dirty, polluted milk.

Pasteurisation

Pasteurisation is the process of warming milk to 63°C for 30 minutes and then heating at 71°C for 15 seconds. This will kill all pathogenic micro-organisms, but some heat resistant bacteria will survive. In addition to keeping milk clean, an aid to keeping milk fresh is cooling it to 5°C immediately after milking or pasteurisation.

Chemical preservatives can be added to milk, but the addition of preservatives is *not* usually recommended. Hydrogen peroxide (H_2O_2) is used in some countries and is considered to be the safest preservative. Recommended levels of inclusion of hydrogen peroxide in milk are 0.01–0.08%, at which levels bacterial growth is inhibited and the milk stays fresh for longer. Treatment slightly decreases the Vitamin C content.

Sterilisation

All organisms can be killed by sterilisation. This is carried out by placing the milk is suitably sized and sealed containers and then raising the temperature to 110–120°C for 20–40 minutes. The milk in most of the sealed packets or containers will keep for long periods under non-refrigerated conditions.

Ultra-high temperature treatment (UHT milk)

Milk raised to the temperature of 150°C for a few seconds is also sterile and the denaturing processes associated with the longer holding time of sterilisation, are reduced. The temperature is raised in special heat exchangers before packaging. It is therefore important to place the UHT milk aseptically into sterilised containers. UHT milk has a shelf-life of up to six months in sealed cartons, non-refrigerated.

Milk products

Milk products include fresh milk and all other products derived from fresh milk. Milk contains more than 80% water, which is expensive to transport. A range of products exist which are valuable because they are usually easier to preserve and store than fresh milk, add value to milk and make transport easier. For example, 36 Indian milk products exist. Such products can be grouped under the following headings:

- fermented milk products
- products from coagulating milk
- butter and ghee
- products from dehydrating milk.

Fermented milk products

Fermentation of lactose by bacteria results in **lactic acid souring**, which is the basis of the manufacture of many cultured dairy products. Under normal storage conditions in the tropics, milk sours in 4–5 hours. The souring has the advantages that it retards the growth of undesirable organisms and makes separation of fat easier.

Maziwa Lala

Maziwa Lala is a fermented milk product made by Maasai pastoralists. Milk is placed in gourds and allowed to ferment for two days to increase the acidity (the acidity kills micro-organisms) and produce a firm coagulum. Slightly fermented milk, with a less acid pH, is given to children. The gourds are sterilised with ashes and washed after making two batches of *Maziwa Lala*. For the commercial production of packaged fermented milk, it is recommended using skim milk warmed to 85°C, cooled to 21–22°C and inoculated with 1.0–1.5% of a special microbiological starter. The product is then ready in 16–20 hours if the temperature is correct.

Milk products are produced in Iran by the Bedouin people who milk sheep and goats then make a number of products for consumption and sale in the nearest towns. *Laban* is a yoghurt-like fermented product, similar to *Maziwa Lala*, which has to be consumed within a day or two of manufacture. *Labneh* is *Laban* in a concentrated form, and can be kept in oil for months. In Nigeria, the Fulani prepare skimmed soured milk (*Nono*) for sale in markets. They may sell this with a prepared cereal food (*Fura*) made from millet.

Yoghurt

Yoghurt is a soured (fermented) product made from boiled whole milk seeded with a suitable culture. The milk should be sterilised at 95°C for 30 minutes to kill pathogens, cooled, a bacterial starter added (*Streptococcus thermophilus* and *Lactobacillus bulgaricus*) and the milk incubated at 38°C for 4–6 hours.

Products from coagulating milk

Cheese

When the casein (protein fraction) in milk coagulates and the water is separated the solids become cheese, in which the nutrients are concentrated and are preserved for a long time. The lower the water content in cheese, the longer the nutrients will be preserved. In soft cheese making, the coagulation is caused by lactic acid produced by bacteria naturally present in the milk. In hard cheese, coagulation is usually caused by addition of **rennet**, a natural product containing the enzyme rennin.

In temperate regions, cheeses of many varieties are made. France and the UK produce many varieties with regional flavours and names. Milk in these countries is produced in the wetter western parts and before the development of reliable transport systems, cheese was made from milk surplus to local requirements. This could then be transported at leisure to the main urban centres and markets.

In the tropics, cheese is not commonly made, because environmental conditions do not favour slow maturation (hard cheese needs several weeks to mature). However, there are numerous examples of cheese production and cheese manufacturing industries have developed in Kenya, Costa Rica, Sri Lanka and India. In Iran, many local cheeses and other milk products are made. Most of the cheeses are produced from pure milk curd or from curd mixed with spices, crushed seeds or other ingredients and are kept in oil or wrapped in leaves for consumption in the winter season. In Sudan, a cheese referred to simply as white cheese is made in a number of centres.

Curds and whey

The simplest form of cheese making is by curdling or coagulating the protein (casein) part of milk to leave whey which contains the milk sugars (lactose) and minerals.

To make curds and whey, milk is warmed to 70°C, cooled to 37°C then rennet is added and the milk coagulates, to form curd, in 30 minutes. If rennet is not available, some of the previous day's whey can be added. The curd is cut with a knife to release the whey, then placed into a wooden mould lined with cotton cloth so that the whey can drain through and be collected. The next day the curd should be placed in brine to preserve it. The whey can be drunk immediately and the curd eaten. If the curd is left for two weeks it develops a more solid consistency, like cheese, and can be eaten. The curds are easier to transport than fresh or fermented milk and keep for longer.

Paneer

In India a similar, simple form of cheese is known as paneer. It is made from buffalo's milk which is heated to 85–90°C for 5 minutes and then cooled to 70°C. Citric acid solution (1%) heated to 70°C is slowly added with continuous agitation until clear whey separates out. When coagulation is complete, stirring is stopped and the curd is allowed to settle. The whey is drained out and the curd scooped into a mould lined with cheese cloth. The curd is wrapped in cheese cloth and pressed for about 10 minutes by applying a pressure of 50 g/cm^2. The pressed paneer is then removed and immersed in clean water chilled to 5–10°C. Deep chilling to a low temperature is essential for a long shelf-life. Paneer can be packaged after being dipped in 5% brine solution. The product is ready for eating, but will keep for a few weeks if refrigerated.

Hard cheese

The most famous hard cheese in the world is Cheddar cheese, originating as a local English cheese from Cheddar in the County of Somerset. The cheese has been manufactured in many parts of the world and the following description of the process is for Cheddar cheese made from buffalo milk at Karnal, India.

Pasteurised milk is inoculated with a starter culture of lactic acid bacteria and then held at 8–10°C for about 12 hours before transfer to a sterilised cheese vat for processing. The temperature of the milk in the vat is raised to 34–35°C and calcium chloride solution (40%) is added at a rate of 15 ml:100 litres milk. More starter culture is then added at a rate of 1.5–2%; as the milk ferments the acidity increases. Rennet is then added and allowed to set for 30 minutes. The curd is cut into cubes (size about 1 cm^3) and cooked for 30 minutes at 39°C. A slow rise in temperature is maintained for a further 20 minutes with an accompanying rise in acidity. The cooked curd is then pressed for 8–10 hours at 35°C. The whey is drained off and the curd cut into long pieces and passed through a milling machine to produce smaller cubes which are washed with hot water for 5 minutes and then placed into moulds. The curd is pressed under its own weight for 30 minutes then with medium pressure for 2 hours, followed by piercing to release air and pressing for 4 hours, then turning and the final pressing for 6 hours. After removal from the press the block cheese is smeared with salt (sodium chloride) and left in a cold store for 2 days. The cheese is then immersed in brine (18%) for 12–15 days, after which time the blocks are removed and allowed to dry on shelves for 2–3 weeks. Finally the blocks of cheese are washed, dried and left for a further ripening (maturation) period of 4–5 weeks.

Butter and ghee

Butter
An emulsion of water in oil, butter contains 80% fat, 16% water, 2% salt and 2% other solids. It is usually made from the cream removed from fresh milk by gravitational separation, surface skimming or centrifugal separation. After separation, the semi-solid butter granules are kneaded into a mass and the water and salt contents adjusted to desired levels.

Butter can also be made from soured milk. In Ethiopia, butter (*kibe*) is made from soured milk (*irgo*) and not from cream. The soured milk is placed in a clay churn or gourd which has been smoked to kill micro-organisms and to add flavour to the product. The churn is stoppered and then agitated, usually by rocking back and forth, until the break point is reached which can be detected by a change in the sound of the milk. The fat globules coalesce into larger globules. The churn is then opened, the butter skimmed off, kneaded into cold water and washed. Butter is made by similar processes in other areas.

Fig 8.1 *Fulani woman with butter for sale in a local market*

Ghee

Also known as clarified butter oil, ghee can be prepared from butter or cream by direct heating to 110–120°C. Ghee can be stored more easily than butter in the absence of refrigeration.

In Iran, a product known as *Kashta* is made by boiling milk and then removing the upper layer of milk fat to leave skim milk. The fat is either consumed as it is (*Kashta*) or is boiled in a pot to evaporate the water to produce ghee (*Samne*). Ghee can be kept in sealed containers for a year or more. In the same way, in southern Darfur, Sudan, milk which is surplus to household requirements is made into ghee (*seme*), or it may be sold to centralised buying points to be made into cheese.

Other milk products

A number of dehydrated and condensed products exist including dried ice-cream mix, dried cheese spread, dried milk (from whole milk or skim-milk), condensed milk and evaporated milk. Removing water by condensing and drying allows these product to be less perishable (increased self-life).

9 Development of milk production

Generalisations about the best policy for increasing milk production are difficult, both because of the variety of systems and the different technical and economic constraints which affect them. Poor planning and the setting of unrealistic targets have contributed to past failures. Often the financial aspects of the operations have not been properly assessed and the level of management expertise required to meet targets not appreciated. On the positive side, much has been learnt from the experiences of past development initiatives and it is now possible to make more rational policy decisions about appropriate ways to develop the industry and the likelihood of success.

In many areas, the growth of urban population and the building of roads and other communication networks has created realisable markets for the sale of milk.

> *The benefits of milk production include weekly income from sales, improved cash flow, the creation of jobs for family members, the utilisation of labour with a low opportunity cost, the use of local materials and crafts, the production of dung and urine for manure and the complementarity of dairying with other farm activities.*

The importance of a realistic milk price for the producer cannot be over-emphasised. Often the milk price is set by the government to suit urban consumers, rather than rural producers. If prices are set, they should be fair for the producer and reviewed annually. If monopolies are given to certain organisations for milk purchase, a fair price is essential, otherwise producers will sell milk to other markets.

Development usually builds upon systems already in operation in the area. Planners choose the best option based on feasibility studies and project proposals put forward by suitably qualified experts. It is impossible to recommend an exact procedure for the development of an industry, but initial consideration should be given to the following issues:

- The potential benefits for the nation and for rural areas.
- Examples of similar development initiatives in other countries.
- Principles of sustainable dairy development and selecting appropriate levels of technology.
- Project identification and formulation, the objectives of development, inputs required and anticipated outputs.
- Strategies for implementation including government and/or agency interventions, producers' initiatives, pricing policies and tariffs.
- Linkages between institutions, such as ministries, banks, donor agencies, universities and training institutions.
- The role of dairy farmer associations.
- Methods of milk marketing, including producer retailers, farmer cooperatives, private sector dairies and state monopolies.
- Milk processing, including the introduction of new products.
- Legislation, quality control and the setting of dairy standards.

Feasibility

Development initiatives will only succeed if the plans are technically and economically, as well as socially, politically and environmentally feasible. Many ideas, although technically feasible, are not feasible in other respects. Social factors must be considered, but people can be persuaded about the viability of change contrary to their previous views. Political constraints might exist, but these can also be overcome. The economics of land use for milk production compared with crops should be considered; the land area required to produce say 2 000 litres of milk a day is relatively small, but maize, rice or other staple and/or cash crops might be better options.

Many past schemes have not succeeded because a false economic environment had been placed around the development project. Funding organisations often provide inputs which otherwise could not be afforded. As a result, when the funding organisation removes its support, the development initiative loses momentum.

Schemes will have a better chance of succeeding if the following criteria apply:

- Increased milk production is technically feasible (i.e. climate and rainfall are good; dairy animals, land and feed are available).
- A suitable and predictable market is available.
- Some financial incentives exist for farmers to produce more milk.

If milk production is considered to be feasible and the above criteria for production are met, then a number of options exist for development and extension efforts. These can be aimed towards:

135

1 helping cattle owners who already produce milk to improve their methods and levels of production (i.e. to provide support for local milk producers such as smallholders or pastoralists),

2 towards expanding the number of milk producers,

3 towards creating a dairy industry where previously there was no milk production.

Each approach involves increasingly greater inputs and problems associated with production. Milk production from extensive dual-purpose beef ranching systems can be considered, as well as creating milk collection schemes, establishing small-scale dairy development schemes or stimulating large-scale dairying.

Development of smallholder milk schemes

In general, smallholder systems are recommended which utilise indigenous breeds and produce milk at moderate levels while producing a weaned calf from the cow. It is likely that in the initial stages of development, production will be based on plots of sown pasture species, such as Napier grass, supplemented with high protein crops, such as leucaena. High energy source perennial crops, such as sugar cane, sweet potatoes, cassava and bananas, could be used to feed dairy animals if the milk price allowed this.

If it has been determined that increased smallholder milk production is technically, socially and economically feasible, then there are a number of stages in the development planning process.

1 Selection of the area

Suitable areas for milk schemes are those where there is a tradition of cattle keeping, a tradition of consuming fresh or soured milk, adequate rainfall and ground water to sustain settled farming, natural pasture available to sustain production and a sizeable market within 20–30 km.

2 Selection of the farmers

The target farming group (i.e. the farmers who would be suitable for a dairy development scheme) would be:

1 Full-time farmers with experience of cattle production and owning cattle.

2 Older people (possibly) who are more likely to make the necessary commitment to the scheme and its success.

3 Farmers who have responded well to extension advice in the past.

Younger people are more likely to leave the area to look for non-farming jobs. Care should be taken to avoid farmers who are just looking for subsidised goods and stock. Ideally a group of farmers should be selected within a 10 km radius, to facilitate collection and to ensure adequate milk supplies to the collection centre.

3 Setting government responsibilities

When a government is planning to implement a small-scale dairy scheme, a level of production and technology should be chosen which can be maintained with available inputs and supplies. Farmers will require continued advice, extension inputs and training if they are to continue to maintain or expand their enterprise. The importance of the price of milk has been mentioned and it is the government's responsibility to provide the legislation to ensure a good price. Some material assistance would be required to provide veterinary services, AI facilities and pasture seeds and cuttings. Assistance with the building of the collection centre and provision of equipment and transport might also be provided from central funds, but these could equally be provided by a farmers' association or an outside donor.

4 Implementation

Interested farmers may begin enterprises by using local animals under natural conditions, with unimproved food supplies and health care. If the environment is improved then production will increase. Improvement can be achieved by:

- reducing climatic hazards
- controlling disease (in particular by vaccination, elimating tick vectors and deworming)
- improving feeding of stock (calves, heifers and lactating cows)
- extending grazing period (to 24 hours if possible).

All schemes run into problems, but many of these can be anticipated and as such, become part of the expected output of the scheme and require forward planning to allow immediate and programmed solutions. For example, it can be assumed that the stock will have low genetic potential and that problems will arise with feeding, health and fertility. The Extension Services should anticipate these problems and have a programme of assistance ready. Advice on water supplies, shade, housing, roof construction, the alleviation of stress, quality control for milk and dairy hygiene should all be available from the start of the scheme. These inputs should be phased with farmers' needs and inputs of their own.

The following table of events indicates some of the processes that should be carried out by farmers:

- Vaccinate against major diseases (rinderpest, anthrax, blackleg, haemorrhagic septicaemia and brucellosis) as necessary.
- Dose cows twice yearly against liver fluke and worms.
- Dip, spray or wash cows weekly with acaricide.
- Fence land if possible.
- Provide water (dam, stream or piped supply).
- Improve pasture.
- Make sure the bull is free from disease.

5   Establishment of milk collection centres

When planning the construction of a milk collecting centre, developers must consider the following factors:

- Daily throughput of milk required to sustain the centre.
- Likely seasonal supply and demand for milk.
- Number of cows needed to produce the milk required.
- Number of farms/owners on which the cows will be based.
- Distribution of these farms/owners.
- Distance of the farm/owners from the collection centre.

It may be considered that a throughput of 2000 litres/day is acceptable. If a cow produces on average 4 litres/day of milk and requires 0.5 ha land, this would require a total of 250 ha (2.5 km²) of grassland, which would produce approximately 0.5 million litres/year of milk for an average 250-day lactation. At one extreme this could be achieved by a small number of intensive dairy farms, each having 50–100 cows. Alternatively this amount of milk could be produced by 100 farmers each having 5 cows. If this were the case, it is likely that these 100 owners would be spread over an area as large as 25–50 km² or greater. The distribution and distance of each owner from the collection centre have implications for the method of collection and the viability of the proposed scheme.

6   Farmers' associations

Advantages can be gained from farmers working together. If dairy farmers' associations can be created this will strengthen the ability of farmers to buy and sell commodities, as well as create a corporate identity and boost morale.

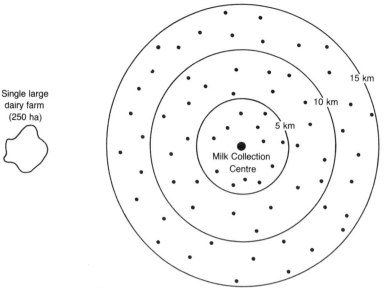

Fig 9.1 *Comparison of a single dairy farm and many scattered smaller farms*

Development of milk production from dual-purpose cattle

Wherever there are beef ranches, there is the potential to produce milk. Ranches are often in remote rural areas, and so access to a market for milk products is necessary. Technical problems of rearing cattle should not be a problem and milk production would require a modification of the existing enterprise.

Ranches could be partitioned to allow the maintenance and milking of small herds of better yielding cows. The ranch can be divided into pockets for suitable enterprises, based on ecological potential and accessibility within the ranch. This is a form of stratification of land resources, to allow better overall utilisation and allows the matching of stock to the production capability of the land. Suitable areas of the ranch for milk production should have some good grassland, river valleys where there is available fodder and a permanent water source. Cows can be upgraded with a dual-purpose bull, then the better female calves used as replacements and the other female and male calves reared for beef. Activities, such as dipping, annual vaccination and supplementary feeding, could be shared with the rest of the ranch.

Development of large-scale dairy production

Large-scale dairy farming requires not only a consideration of the technical problems associated with producing milk, but also the problems of managing a complex business enterprise.

The animal husbandry requirements are much the same as for other forms of milk production, but require more attention to detail. In intensive production it is more likely that high yielding breeds will be used in order to maximise milk yield to cover the high fixed costs. High yielding animals are more likely to suffer from stress, nutritional deficiencies and health problems than lower yielding, adapted local breeds. This requires that the management and husbandry capability of the staff is of a high calibre. People, well trained and experienced with stock, milking and dairy husbandry are required. The farm managers should be good agricultural graduates with five or more years experience of dairy farming. Good farm records must be kept and constant attention given to health, fertility, housing defects, pasture management and staff problems.

Glossary

Abomasum Fourth compartment of the ruminant stomach; the true glandular stomach.

Acaricide Diluted chemical preparation used in dip-tanks or hand-sprays to kill ticks (acarids) on animals.

Adaptation The development of characteristics which improve the chance of survival in a given environment.

Ab lib. Providing free access to food.

Ambient temperature Temperature of the surrounding air.

Anaemia Deficiency in haemoglobin, often accompanied by a reduced number of red blood cells, causing paleness, weakness and breathlessness.

Anoestrus Absence of oestrus; a period of sexual inactivity.

Anthelmintic Chemical used to kill worms in the digestive tract.

Artificial insemination (AI) Insemination of the cow artifically by a skilled technician using collected semen and special equipment.

Body condition Fatness, determined by condition scoring.

Browse Edible parts of trees, bushes and other woody plants (mainly leaves, twigs and fruits) which are available for animal consumption.

By-product Any part of a crop, other than the main harvested product, which can be used for animal feed.

Carrying capacity The number of stock that can be supported, either year round or seasonally, over a long period (expressed in hectares per livestock unit).

Chronic Describes a disease continuing, often at a low level, for a long time.

Colostrum The first milk produced by a female for her offspring to supply the young's initial requirements of nutrients, Vitamins A and D and protective antibodies.

Concentrate feed Ruminant food made from combinations of good quality constituents, such as oil seed residues, molasses and mineral additives.

Conception Union of the egg and sperm; fertilisation.

Constraint A factor which impedes the process of change, development or uptake of new ideas; usually grouped under the headings social, economic, technical, political or environmental.

Corpus luteum The yellow body formed in the ovary from the Graafian follicle (after release of the ovum), under the control of luteinising hormone. After fertilisation, it secretes progesterone hormone to maintain pregnancy.

Corral (kraal) A fenced enclosure for stock.

Crossbreeding Mating animals of different breeds.

Criollo cattle Term for Local *Bos taurus* cattle (of European origin) in Latin America.

Cull To remove unwanted animals (e.g. non-breeding cows or aged stock) for sale or slaughter.

Deferred grazing Simplest form of fodder conservation, by closing off an area for the latter half of the wet season and allowing the pasture to grow into standing hay.

Dry matter (DM) Part of a food remaining after all the water has been removed from it by drying.

Dry period Period of non-lactation between two periods of lactation.

Dual-purpose Cattle which have two primary uses, such as for draught power and milk production or for milk and beef production.

Ectoparasites Parasites which live on the skin of the animal, such as ticks, mites, lice and fleas.

Embryo The developing young animal, from conception to three months, in cattle.

Endemic A disease usually existing in an area.

Endoparasites Parasites which live inside the animal's body.

Extensive Systems which use large area of land per animal unit.

Fodder Any bulky green or dried plant material used to feed stock.

Fodder crops Cultivated annual or perennial crops grown principally for harvesting and feeding to animals in a fresh or dried state.

Fetus The developing young animal, from three months onwards, in cattle.

Food supplement Food which is given in small amounts to complement a roughage diet.

Forage Vegetation available as food for livestock or game.

Forage crops Cultivated crops either grazed or browsed directly.

Genetic capacity Potential of an animal to produce, as defined by its genetic make-up.

Gestation Period during which the embryo and fetus grow in the uterus of the mother; cattle = 40 weeks.

Globulin One of the groups of protein present in blood plasma; gamma immunoglobulins associated with immunity and resistance to disease.

Heat detection   The ability to detect animals which are in oestrus.

Heritability (h²)   The proportion of variation in a trait which results from genetic effects.

Husbandry   The day to day practice of looking after farm animals.

Hybrid vigour   Increased vigour (in terms of growth, fertility and production) in a cross between genetically different lines when compared with the same characteristics in either parental line.

Immune   Protected against specific disease.

Inbreeding   Mating closely related animals.

Indigenous   Originating in and native to a particular region.

Intensive   Agricultural systems which usually use small areas of land, but high inputs of capital.

Joule (J)   International unit of energy (1 cal (obsolete unit) = 4.2 J).

Lactose   The sugar in milk.

Libido   Sex drive; usually refers to male animals.

Livestock unit   Standardised animal unit to which different ages, types or species of livestock can be related for purposes of matching forage availability to animal needs.

Luteinising hormone (LH)   Secreted by the anterior lobe of the pituitary and normally initiates the formation of the corpus luteum in the female and the secretion of testosterone in the male.

Maintenance   Term used to describe the nutrients needed to keep the animal alive, but with no production.

Metabolic weight   Liveweight raised to the power 0.75 ($LW^{0.75}$).

Metabolism   Life sustaining processes in the body, including nutrition, energy production and growth.

Oestrus   Recurrent period of receptivity of females to males, resulting from the regular series of hormonal events known as the oestrous cycle.

Pastoralist   A person who looks after ruminant livestock as a means of subsistence, usually in dry areas where other forms of agricultural crop production are marginal because of low and unpredictable rainfall.

Pregnancy diagnosis   Determination by manual examination of the cow's internal genitalia whether or not she is in calf.

Productivity   Estimate of the efficiency of production; measures output in terms of inputs made by the producers.

Puberty   Growing stage during which the reproductive system acquires its mature form and function.

Rumen   First stomach of the ruminant animal in which food is fermented by bacteria, protozoa and fungi before final digestion in the abomasum and lower digestive tract.

Semen   Fluid, produced by the male reproductive organs, containing sperm.

Service   Mating and fertilisation of the female animal artificially or by a male animal.

Smallholder   A settled farmer who controls the farming activities on a small plot of land; often involved in mixed-farming.

Stocking rate   The intensity at which an area is stocked, expressed as the number of head or livestock units per square kilometer.

Stress   State in which the animal's homeostatic and behavioural balance is disturbed by disease, climate, physical environmental conditions or management factors, so that the animal does not produce optimally.

Subsistence   Production of food to meet the family's day-to-day needs throughout the year.

System   A series of inter-related components which operate together to produce an output.

Transhumance   Seasonal movement of people and their livestock.

Vasectomise   Cutting the vas deferens (sperm duct) so that sperm cannot leave the male body.

Venereal disease   Disease known to be transmitted during mating (e.g. brucellosis).

Weaning   The process of changing the diet of a young animal from milk to solid food.

Bibliography

Agricultural Research Council *The Nutrient Requirements of Ruminant Livestock*. Commonwealth Agricultural Bureau, Farnham Royal, Slough, UK. (1980) pp.1–347

Agricultural Research Council *The Nutrient Requirements of Ruminant Livestock. Supplement No. 1*. Commonwealth Agricultural Bureau, Farnham Royal, Slough, UK. (1984) pp.1–45

Ansell, R.H. Extreme heat stress in dairy cattle and its alleviation: A Case Report. In: Clark, J.A. (Ed) *Environmental Aspects of Housing and Animal Production*. Butterworths, London, UK. (1981)

Baker, R. Stages in the development of a dairy industry in Bunyoro, Western Uganda. In: *Pastoralism in Progress: Readings on the development of traditional cattle herding areas in Africa*. Development Study 6, 1975. University of East Anglia, Norwich, UK. (1975)

Barret, M.A. and Larkin, P.J. *Milk and Beef Production in the Tropics*. Oxford University Press, Oxford, UK. (1970)

Bayer, W. Agropastoral herding practice and grazing behaviour of cattle in the subhumid zone of Nigeria. *ILCA Bulletin* No. 24. International Livestock Centre for Africa (ILCA), Addis Ababa, Ethiopia. (1986)

Coughenour, M.B. *et al*. Energy extraction and use in a nomadic pastoral system. *Science* (1985) 230 pp.619–625

De Leeuw, P.N., Bekure, S. and Grandin, B.E. Aspects of livestock productivity in Maasai group ranches in Kenya. *ILCA Bulletin* No. 19 pp.17–20, ILCA, Addis Ababa, Ethiopia. (1984)

FAO *Manual for Animal Health Personel*. Food and Agriculture Organisation, Rome, Italy. (1983)

FAO *Production Yearbook*. Food and Agriculture Organisation, Rome, Italy. (1988) 42

Felius, M. *Genus* Bos: *Cattle Breeds of the World*. Rahway, USA. MSDAGVET. (1985) pp.234

Humphreys, L.R. *Tropical Pasture and Fodder Crops*. (2nd edition) Longman, London, UK. (1987)

Gatenby, R.M. Shelter for Animals in Hot Climates. In: Grace, J. (Ed) *Effects of Shelter on the Physiology of Plants and Animals*. Swets and Zeit Linger, Lisse, Netherlands. (1985)

145

Goldson, J.R. and Ndeda, J.O. Cattle in Kenya – 1 Milk Production. *Span: Progress in Agriculture* (1985) 28 pp.111–113

Knudsen, P.B. and Sohael, A.S. The Vom herd: A study of the performance of a mixed Friesian/Zebu herd in a tropical environment. *Tropical Agriculture* (1970) 47 pp.189–203

Lowman, B.G., Scott, N.A. and Somerville, S.H. *Condition scoring of cattle. Bulletin 6.* East of Scotland College of Agriculture, Scotland, UK. (1976)

Macfarlane, J.S. and Worral, K. Observations on the occurrence of puberty in *Bos indicus* heifers. *East African Agriculture and Forestry Journal* (1970) 5 pp.409–410

Mason, I.L. *World Dictionary of Livestock Breeds.* (3rd edition) Commonwealth Agricultural Bureau International, Wallingford, UK. (1988)

McDonald, P., Edwards, R.A. and Greenhalgh, J.F.D. *Animal Nutrition.* (4th edition) Longman, London, UK. (1987)

McDowell, L.R. *Vitamins and Animal Nutrition: Comparative Aspects to Human Nutrition.* Academic Press Inc., San Diego, USA. (1989)

McIlroy, R.J. *An introduction to tropical grassland husbandry.* (2nd edition) Oxford University Press, Oxford, UK. (1972)

Maule, J.P. *The Cattle of the Tropics.* University of Edinburgh Press, Edinburgh, UK. (1990)

Meyn, K. and Wilkins, J.V. Breeding for milk in Kenya, with particular reference to the Sahiwal stud. *World Animal Review* (1974) 11 pp.24–30

Musangi, R.S. *Dairy Husbandry in East Africa.* Longman, Nairobi, Kenya. (1971)

Olayiwole, M.B. Feeding and Management of Dairy Cows at Shika, Zaria. In: Loosli, J.K., Oyenuga, V.A. and Babatunde, G. M. (Eds) *Proceedings of an International Symposium on Animal Production in the Tropics, held at Ibadan, 26–29 March 1973.* Institute of Agriculture, Ibadan, Nigeria. (1974)

Oliver, J. *Dairy Farmers Handbook.* National Association of Dairy Farmers of Zimbabwe, Harare. (1987)

Peters, A.R. and Ball, P.J.H. *Reproduction in Cattle.* Butterworths, London UK. (1987)

Pullan, N.B. Condition scoring of White Fulani cattle. *Tropical Animal Health and Production* (1978) 10 pp.118–120.

Roy, J.H.B. *The Calf.* Butterworths, London, UK. (1980)

Russell, K. *The Principles of Dairy Farming.* (8th edition) Farmers Press Ipswich, UK. (1988)

Schillhorn van Veen, T.W. and Loeffler, I.K. Mineral deficiency in ruminants in subsaharan Africa: A Review. *Tropical Animal Health and Production* (1990) 22 pp.197–205

Schmidt, G.H. *Biology of Lactation.* W.H. Freeman, San Fransisco, USA (1971)

Sewell, M.M.H. and Brocklesby, D.W. (Eds) *Handbook on Animal Disease in the Tropics.* (4th edition) Bailliere Tindall, London, UK. (1990)

Smith, A.J. *Milk Production in Developing Countries.* University of Edinburgh Press, Edinburgh, UK. (1985)

Spedding, C.R.S. *The Biology of Agricultural Systems.* Academic Press, London, UK. (1975)

Thuraisingham, S. *Proposals for large-scale dairy development in west Malaysia.* Veterinary Services Department, Kuala Lumpur, Malaysia. (1969)

Turton, J.D. Progress in the development and exploitation of new breeds of dairy cattle in the tropics. In: Smith, A.J. (Ed) *Milk Production in Developing Countries.* Edinburgh University Press, Edinburgh, UK. (1985)

Index